SPACE TECHNOLOGY VETERANS

QUINTESSENCE OF NANO-SATELLITE TECHNOLOGY

SMALL IS BIG

I0480058

PLANET AEROSPACE INDIA

INDIA · SINGAPORE · MALAYSIA

Notion Press

No. 8, 3rd Cross Street
CIT Colony, Mylapore
Chennai, Tamil Nadu – 600004

First Published by Notion Press 2020
Copyright © Planet Aerospace India 2020
All Rights Reserved.

ISBN 978-1-64951-662-6

This Book is dedicated to

Padma Vibhushan

Prof. U.R. Rao

Father of the ISRO Satellite Program

CONTENTS

"The development of the nation is intimately linked with understanding and application of science and technology by its people."

Dr. Vikram Sarabhai,
Founding father of Indian Space Program

"To succeed in your Mission you should have single minded devotion to your goal."

Dr. Abdul Kalam,
Former President of India

"While Science & Technology can be powerful tools for National development, they need to be grown, handled and applied with care to better the lives of the people."

Prof. Satish Dhawan,
Former Chairman, ISRO

FOREWORD

As human evolution on 'planet earth' continues, the untiring efforts of humans to grapple with issues of survival on 'planet earth' has led them to constantly develop new tools for understanding the complex nature in which they live and also improve their ability to survive on 'planet earth'. In the quest for their understanding of the ways of the 'planet earth', they found it necessary to go beyond the earth and establish platforms orbiting the earth. Using these platforms, humans have been able to improve their ability to communicate amongst themselves across the entire globe, broadcast signals across the globe, found ways of locating themselves with simple gadgets using navigation signals, improved their ability to deal with cyclones, hurricanes and tsunamis as well as improve their knowledge of the weather. The use of space technology-based tools has become so much an integral part of human life on the earth that it is impossible to visualise life without the support of such tools. These platforms are being used for exploring the universe and improve our understanding of the cosmos.

Even as the development of space technology is changing gears from government-dominated activities to commercial entrepreneur-driven activities to push ahead in dealing with the Fourth Frontier "space" to grapple with issues of space exploration, space exploitation, space adventure, space tourism and space-enabled services for improving the quality of life on 'planet earth', the need for innovation and breakthroughs in technology are essential. This can happen only when students, who are the future citizens of the 'planet earth', work on pushing the frontiers of technology to address various issues of sustainable living on 'planet earth'.

The process of making the next generation of enterprising student community aware of the technology of satellites and Nanosatellites, in particular, is the effort of the authors of this book. This book is a one-stop-shop for information and I am very certain that both the students and teachers will find it very useful in their pursuit of Nanosatellite technology development.

A.S. Kiran Kumar
Former Chairman, ISRO/Secretary, DoS

PREFACE

When *"Small Spacecraft Technology, State-of-the-Art"* report was first published by NASA in 2013, around 250 CubeSats and 100 other small satellites (under 50 kg class) were already launched. Since then, over 300 CubeSats and nearly 350 small satellites have been launched in 2017 alone and these numbers have been ever increasing. This provided a lot of excitement and enthusiasm among young entrepreneurs as well as the student community to venture into space science and technology. Many universities/academic institutions in India and other countries have taken up small satellite projects to understand the challenges involved in building and launching satellites. Planet Aerospace was approached by many such institutions to organise workshops/seminars to provide basic training on Nanosatellites. During one such initiative, Planet Aerospace experts felt the need to publish a book on Nanosatellite covering all aspects of design, development and testing of a Nanosatellite. This book is written by domain experts, who have spent more than three decades of working in their respective areas of satellite systems. Even though many books are available in the market that provide theoretical aspects, several intricacies and nuances in design, analysis, development and testing are covered in this book.

This book is structured such that each chapter provides detailed design aspects related to each sub-system of a Nanosatellite, including design options and trade-off analysis. After providing a broad introduction to the topic of small satellites, the first chapter provides an in-depth overview of how to plan and operate a typical Nanosatellite mission along with choosing the type of orbit. The second chapter on payloads gives an idea on some of the options available for the selection of a payload keeping in mind the constraints of weight, volume and power of a Nanosatellite.

The next three chapters describe mechanical systems of a Nanosatellite such as structure, thermal control and deployment mechanisms highlighting the pros and cons of design options along with the criteria for selections of various materials.

Subsequent chapters describe details of electrical systems of a Nanosatellite such as power, digital systems, onboard computer, RF communication systems, including the ground station. Even though design concepts are mostly the same for both big and small satellites, the challenge lies mainly in optimising the system design within the constraints of a Nanosatellite.

The next chapter on attitude determination and control system highlights the control aspects of a Nanosatellite. It starts from fundamental theory and provides an in-depth study of algorithms based on attitude sensors and control elements. The criticality of controlling the attitude errors for different missions is also explained.

The chapter on assembly, integration and testing (AIT) gives a fairly good idea of all activities of a Nanosatellite encompassing the development phase, satellite assembly and integrated satellite testing as well as pre-launch phase. Though there may be subtle differences in AIT operations between a big and a Nanosatellite, basic principles and methodologies remain the same.

Last but not the least, a chapter is dedicated to product assurance aspects and guidelines to be followed during all phases of a satellite's development to ensure a high probability of a mission's success.

Appendices have been added at the end of the book giving more details on specific topics for providing a better understanding to readers. Various acronyms used in this book are summarised at the beginning of the book to serve as ready reference.

It is our opinion this book can be used as a textbook for students pursuing a course on Satellite Technology. It will also enhance their knowledge and provide pleasure to those who have an aptitude for space technology in general and satellite systems in particular

Suggestions and contributions have also been received during the review process of the draft of this book from numerous people. The content in this book is not intended to be exhaustive – no such assessment can be given based on the pace of the technology development in this area. New technology is developing continuously and emerging technologies will mature to become the state-of-the-art in the future.

NOTE:

We will endeavour to update this book with new technologies in the area of Nanosatellites in future editions. We seek your feedback on the contents of this book and please contact PLANET AEROSPACE for the same.

R.K. Rajangam
President
Planet Aerospace

ACKNOWLEDGEMENTS

Planet Aerospace, a Registered Association formed by former Scientists/ Engineers/Administrators of ISRO conducted many seminars/workshops in University Engineering Colleges to impart the students with knowledge on basics of Space technology and applications. As a feedback from interactions with the students, an idea was mooted that a concise book on satellite technology in general and Nanosatellites in particular will help not only students but also others who are inclined to learn about this field of technology.

At the outset, we are ever thankful to ISRO, the prestigious Organization in which we had the privilege of serving more than three decades in various important Projects and gained valuable experience in design, development and launching of satellites and associated infrastructure. The contribution of all authors who are experts in the concerned field is acknowledged with gratitude for their dedicated knowledge and time.

An editorial board comprising M.Venkata Rao, L.S. Satyamurthy, A.A. Bokil and R.K Rajangam have gone through the contents of this book several times and brought it to present shape. We are extremely grateful for their sincere efforts in publishing this book. We also thank G.D Murthy for meticulously going through the manuscript and making corrections and valuable suggestions. We acknowledge with gratitude the contributions of H.R.Nagendra, V.N Purohit, S.G Raghavan and C.A Prabhakar for their valuable inputs.

We express our heartfelt gratitude to Dr. K.Sivan, Chairman ISRO/Secretary DOS for providing all the encouragement for publication of this Book.

We are very grateful to Dr. A.S Kiran Kumar, former Chairman ISRO/ Secretary, DoS, who has gone through the book, made valuable suggestions and written a Foreword to this book.

We are immensely thankful to our stalwarts, Dr. K. Kasturi Rangan and Dr. P.S Goel who were kind enough to go through this book and guide us in improving the contents.

We thank G.R. Hathwar who mooted the idea of bringing out this Book on Nanosatellite technology and spearheaded the publication of this book.

We are especially grateful to M.Venkata Rao who was responsible for editorial corrections and proof reading of the book.

Finally, we sincerely thank M/s Notion Press Media Pvt.Ltd,. Chennai for all their efforts in editing, design of cover page and layout, printing and publishing

G. Raghavendra Hathwar
Planet Aerospace

EDITORIAL NOTE

Space Science and Technology… Inflexion point for the ignited minds…
Some years remain etched in the history of mankind
The Year 1969 was one such with several milestones of national and international significance. The mankind took a "Giant Leap" as Astronaut Neil Armstrong became the First man to set foot on Moon and Indian Space research organization (ISRO) started its Space odyssey. If Boeing 747 Aircraft made its maiden passenger flight in America, the first Rajdhani Express train was rolled out in India.

1969 was such an illustrious year when our Indian Space program was formalized on 15th August on the India's Independence Day, under the leadership of the one and the only Dr. Vikram Sarabai. Six years later in 1975 India launched its First Satellite 'Aryabhata', which was built in a record time under the leadership of the doyen Astro Physicist Prof. U.R. Rao, in whose illustrious name the ISRO Satellite Centre is renamed as URSC.

The Satellite was launched by Russian Rocket 'Vostok' from Kapustin Yar cosmodrome in southern Russia and named in the honor of the 4th century Indian Astronomer 'Aryabhata', who enunciated the mathematics of planetary motion. Satellite Aryabhata became the harbinger for the Indian Space program leading to the development of myriad of advanced Satellites, launch vehicles and inter planetary missions, executed under one agency ISRO.

To the young minds, Space is an exciting arena where the dreams manifest into thoughts, which stimulates the action, leading to a discovery or an invention. The unending and unraveling mystery of cosmos is the beginning for a student to channelize his intellect in nurturing the thoughts and knowledge through the frontiers of space technology, which is truly the inflection point for the ignited minds.

That is what the Planet Aerospace, which is an organization of Satellite Technology experts from ISRO, has attempted to prepare the next generation space scientists/ technologists to take up the challenges of perpetuating Space arena for greater benefits to one and all. With the aim of imparting the Knowledge to the student community, Planet Aerospace has taken this initiative of opening the chapters of this book "Quintessence of Nano Satellite Technology". You. The readers. Enjoy this exciting new Space scenario.

L.S. Satyamurthy,
Vice President, Planet Aerospace.

LIST OF ANNEXURES

ACRONYMS

ADCS	Attitude determination & control system
AIS	Automatic identification of ships
BER	Bit error rate
CCD	Charge coupled device
CFRP	Carbon fibre-reinforced plastic
CMOS	Complementary metal-oxide semiconductor
CPU	Central processing unit
CVCM	Collected volatile condensable material
DOD	Depth of discharge
EDAC	Error detection and correction
EIRP	Effective/Equivalent isotropic radiated power
EMC	Electromagnetic compatibility
EMI	Electromagnetic interference
EO	Earth observation
ESD	Electro-static discharge
FDI	Failure detection & isolation
FPGA	Field programmable gate array
FSK	Frequency shift keying
GaAs	Gallium arsenide
GEO	Geostationary earth orbit
GFRP	Glass fibre reinforced plastic
HgCdTe	Mercury cadmium telluride
IOT	Internet of Things
ISRO	Indian Space Research Organisation
ITU	International Telecommunications Union
LEO	Low earth orbit
LTP	Lower trip point
MEMS	Micro-electro-mechanical sensors
MEO	Medium earth orbit
MFS	Multifunctional structure

MLI	Multi-layer Insulation
MPPT	Maximum power point tracking
MRR	Mission Readiness Review
MVL	Majority voting logic
MWIR	Middle wave Infra Red
Ni-Cd	Nickel-cadmium
Ni-H2	Nickel Hydrogen
NMI	Non-maskable interrupt
NORAD	North American Aerospace Defence Command
NRT	Near real-time
NRZ-L	Non-return to zero – level
NRZ-S	Non-return to zero – space
OBC	Onboard computer
OBT	Onboard timer
OSR	Optical solar reflector
PA	Product assurance
PCM	Pulse code modulation
POD	Preliminary orbit determination
PPT	Pulsed plasma thruster
PROM	Programmable read-only memory
PSK	Phase shift keying
PSLV	Polar satellite launch vehicle
RAM	Random access memory
RF	Radio frequency
SMA	Shape memory alloy
SNR	Signal-to-noise ratio
SSLV	Small satellite launch vehicle
SSO	Sun-synchronous orbit
STEM	Storable tubular extendable member
SWIR	Short wave infrared
TCS	Thermal control system
TCXO	Temperature controlled crystal oscillator
TEC	Total electron content
TLE	Two-line element
TML	Total mass loss
TTC	Telemetry, tracking and command
UHF	Ultra high frequency
UTP	Upper trip point
VHF	Very high frequency
VLSI	Very large scale integrated circuit

INTRODUCTION

L.S. Satyamurthy

"Nanosatellites are the gateway for a new generation of affordable space technology, applications and services for reaching the unreached"

The introduction provides the much-required information on the genesis of artificial earth satellites, which have become the harbinger of the space technology ecosystem, covering the civilian space domain of launch vehicles, ground stations, satellite control centres, payload data reception centres and the end-user product repository for application areas. The reader will experience this evolutionary technology inversion in the development of a Nanosatellite, leading to a small satellite performing complex functions of a big satellite.

Truly in the Indian space parlance, a satellite is regarded as the celestial goddess of space ushering service to the mother earth. The Nanosatellite enhances this myth to show that 'small is cheaper, faster and performer'.

Enjoy reading every bit of this valuable text.

1.1 THE BEGINNING OF 'SPACE ERA'

In the history of mankind, the dawn of 'space era' began with the launch of the first artificial earth satellite 'Sputnik 1' by the then Soviet Union on October 4, 1957. The Sputnik 1 was followed by a larger satellite, Sputnik 2 in November 1957, which carried a living passenger - a dog named Laika. During the same period, America's first satellite 'Explorer 1' was launched on January 31, 1958.

This momentous event heralded the onset of the major space race of the 1960s leading to a path-breaking military and political competition among some of the scientifically advanced nations for reaping the benefits of space science and technology for civilian and military applications. The size and mass of these satellites were primarily limited by the capabilities of launch vehicles available at that time. As the launch vehicle capabilities advanced, so did the mass and data throughput of the satellites.

1.2 EARLY SPACE EFFORTS

Early space efforts were dominated by competition for political and scientific supremacy through the exploration of the solar system and the human space flight culminating in the Apollo Moon landing mission and gaining greater access to space. The economic exploitation of satellites to provide civil communications for telephone, TV, observation of the earth (EO) for meteorology and land resources planning and utilisation were major takeaways from the space race.

The privileged access to space was the prerogative of a few technically advanced countries, which gave them an overwhelming advantage over the developing world for about two to three decades. However, some of the equally conscientious developing countries began to send their own satellites into space, as the benefits of space technology penetrated through the society for the development of mankind at large. Remote sensing satellites with improved capabilities served several applications on land, ocean and atmospheric studies, tracking changes in forest cover and other details on the earth's surface. Telecommunication satellites brought to the fore, long-distance telephone calls and eventually live television broadcasts from across the world, which has now become an integral part of human life.

1.3 IMPORTANT TRANSFORMATION

The balance of power slowly tilted towards central median with the advent of miniaturisation of electronics, optics and micro-electro-mechanical (MEM) devices that enabled smaller nations physically build smaller satellites with limited budget and resources. Also, the utilisation of launch vehicle capacity to carry secondary payloads or piggybacks on a bigger satellite for a shared launch catalysed the small satellite era in the 1980s. With the miniaturisation of mobile phones, computers and other hardware, it is now possible to send up much smaller satellites into orbit that enable science, telecommunications - voice, video and data, environmental monitoring and other functions.

The renaissance of the Nano-Micro satellite era can be attributed to the fast changing technologies cutting longer gestation periods for practical realisation of the satellite systems. The agencies/industries are responding to the future expectations by building smaller spacecrafts and launching them at shorter time frame and providing products quickly for the users. The exciting era of Nano-Micro satellites began recently a few years ago. In November 2013, NASA launched 29 satellites using Orbital Sciences Minotaur-1 Rocket. Thirty hours later, Kosmotras, a Russian company, carried 33 satellites into a similar orbit using Dneper Rocket. ISRO broke these records by launching 102 small satellites in a single launch of PSLV in 2017.

1.4   WHAT IS NANOSATELLITE?

After detailed deliberations, the international space community broadly classified satellites according to their lift-off mass as shown in the table below. As per this classification, Nanosatellites come under 1 to 10 Kg class, which are most suitable for carrying certain payloads. This is why many students all over the world are developing and launching Nanosatellites for various scientific and technological experiments.

Big	500 Kg
Mini	100 – 500 Kg
Micro	10 – 100 Kg
Nano	1 – 10 Kg
Pico	0.1 – 1 Kg
Femto	< 100 gm

A Nanosatellite can be developed either by adopting a standard bus (a satellite platform) or by designing an own bus. The standard bus can be a 'CubeSat', which was developed in 1999 by California Polytechnic State University (Cal Poly) and Stanford University, the USA. This is a 10 cm x 10 cm x 10 cm cubic shaped structure with a mass of about 1.33 Kg – each CubeSat is referred to as '1U' and such units can be stacked as 2U, 3U etc,. These CubeSats along with sub-systems such as power, onboard computer and attitude control systems are commercially available in the market with standardised interfaces and specifications. However, designing and developing a Nanosatellite by students on their own gives hands-on experience and knowledge.

1.5   GLOBAL ECONOMICS OF A NANOSATELLITE

Global space economics in 2012 was valued at US $ 300 billion in commercial revenue and government budgets are expected to double by 2030. This growth is expected to come from the commercial sector, driven primarily by entrepreneurs and new business ventures. The global small satellite market is valued at $ 600 million to $ 1 billion annually with an estimated 2200 to 2700 launches planned during the 2015-2023 window.

A recent study by a company 'Commercial Space Works' projects strong growth in the Nano and Microsatellites development with an estimated launch of 120 to 200 Microsatellites globally by 2020 as against 33 Microsats during 2013. The development of Nano and Microsatellites continue to be led by the civil sectors but the defence and intelligence community will soon grab their due share of the market.

1.6 DISRUPTIONS IN TECHNOLOGY

In the 20th century, there was a famous saying in techno-business parlance that "by the turn of the century, computers would have a billion words of memory". Further, the doyens of the microprocessor industry (Moore's Law) predicted that the number of transistors on a chip will increase exponentially, doubling every two years or 10 times in 6.5 years. We have seen these predictions come true and the trend is continuing.

The microelectronics revolution in terms of production techniques and processes developed for the consumer market (automobiles, mobile phones, TVs) have reached such a reliable stage that it is safe to assume that random component failures have been virtually eliminated. This is against the earlier methodology and belief that achieving high reliability involves rigorous component/device qualification testing practised by various space agencies/industries. The electronic components, manufactured for the consumer market called "commercial off the shelf (COTS)", have thus effectively set a new trend for enhanced reliability devices for operational satellites. Other disruptive technologies include:

- Miniaturised optics for mobile cameras
- MEMs. devices such as accelerometers, sensors
- High-efficiency solar cells (multi-junction GaAs - gallium arsenide semiconductor)
- High-efficiency batteries (Li-Ion; lithium-ion)
- Advanced lightweight materials

However, while using the COTS, due care should be exercised in terms of adequate study and demonstration of these devices to space radiation and thermal environments experienced in different earth orbits and durable for the intended life of spacecraft.

This is a real challenge for the designers of the spacecraft especially because, these COTS devices come from different manufacturers and manufacturing setups and production batches. The most important thing for the Micro-Nano satellites is that enough safeguards and precautions need to be applied for avoiding catastrophic failures of any Nano-Micro Satellite missions and hence a simple and in built quality control and reliability assurance program is essential without conflicting the cost and schedule.

1.7 THE EVOLUTION OF NANOSATELLITE

The initial small satellites were pioneered by Radio Amateurs (HAM Radio operators), who wanted to get better point-to-point coverage from a satellite which was hitherto not possible because of limitations due to ionospheric disturbances. A group of Radio Amateurs form California, USA built a 10-Kg satellite OSCAR 1 (Orbiting Satellite Carrying Amateur Radio), which was launched as a secondary payload on the Delta Rocket. OSCAR 1 was power-limited because the satellite used a battery that got depleted after three weeks in orbit due to the absence of solar cells for charging the battery.

Surrey University and Surrey Space Technology Limited UK, have been the pioneers in Small satellite technology and one of the prime movers of Nano- Micro Satellite development

Gradually, Small, Micro and Nanosatellites adopted most of the technologies established for larger satellites, especially in the domain areas of thermal engineering, attitude control and stabilisation, digital electronics and onboard computer, optical/inertial sensors and control actuators, active and passive propulsion systems.

It is envisaged that a large proportion of Nano-Micro Satellites are useful for earth observation purposes. The earth observation application is constrained by adequate and repetitive earth imagery data availability among service providers like government or private organizations in the prime areas of agriculture, land resource mapping, weather forecasting and disaster management, forestry and even wild life conservation purposes. Another important area which is getting attention recently have been the benefits of usage of smaller Satellite constellation for Tele-health and Tele-education area ushering low cost connectivity access.

With the advent of CubeSat platform in the 1990s, many universities and industries launched hundreds of Nanosatellites into low earth orbits for various applications such as earth remote sensing, automatic ship identification, atmospheric studies and technology demonstration. A compendium of Nanosatellites launched by different agencies, institutions/universities is given in the website **www.nanosats.eu.**

1.8 ISRO'S SMALL SATELLITE PROGRAMME

With the launch of its first satellite *Aryabhata* on April 19, 1975, ISRO established its space programme, which has grown into a major space agency in the world during the past four decades. Today, ISRO has self-reliance in the areas of launch vehicles (PSLV, GSLV), communication and meteorology, earth observation, navigation and scientific satellites.

With the main objective of introducing the student community to *'Space Science and Technology'*, ISRO initiated its small satellite programme in 2002. ISRO provided cost-free launch for small satellites developed by academic institutions in addition to providing technical support during their development. A list of such student satellites launched so far is given below. ISRO also launched Small and Nanosatellites from various countries on its PSLV with a record of 102 small satellites launched by PSLV-C37 in February 2017. Based on the predicted huge demand for small satellite launch services, ISRO is developing a 'Small Satellite Launch Vehicle' (SSLV).

ISRO, through its official website, offers two options to academic institutions – (i) develop an innovative payload which will be launched by ISRO on its satellite and will also provide post-launch support, (ii) develop a Nanosatellite and ISRO will provide free launch.

Sl. No	Name	Launch Date	Launch Mass	Launch Vehicle
1	Kalamsat-V2	Jan 24, 2019	1.26 Kg	PSLV-C44
2	NIUSAT	Jun 23, 2017	15 Kg	PSLV-C38
3	PISAT	Sep 26, 2016	5.25 Kg	PSLV-C35
4	PRATHAM	Sep 26, 2016	10 Kg	PSLV-C35
5	SWAYAM	Jun 22, 2016	1 Kg	PSLV-C34
6	SATHYABAMASAT	Jun 22, 2016	1.5 Kg	PSLV-C34
7	Jugnu	Oct 12, 2011	3 Kg	PSLV-C18
8	SRMSat	Oct 12, 2011	10.9 Kg	PSLV-C18
9	STUDSAT	Jul 12, 2010	< 1 Kg	PSLV-C15
10	ANUSAT	Apr 20, 2009	40 Kg	PSLV-C12

1.9 NANOSATELLITE - FUTURE SCENARIO

While universities/academic institutions continue to develop and launch Nanosatellites for new scientific research and/or proof of concept technology experiments, recent trends show that major industries are deploying a "constellation" of Nanosatellites for operational services such as the Internet of Things, weather monitoring and rapid observation of the earth's features. For example, Planet Labs, a company in California, launched about 200 Nanosatellites which can provide imagery of any given land area on the earth with a repetivity of a few hours. Boeing Company of the USA is planning a mega constellation of about 1,000 or more Micro-Nanosatellites to be deployed into different orbital planes and operating in higher frequency bands. Similarly, SpaceX and One Web of the USA are planning for a much larger constellation of such satellites.

Owing to this large demand, satellite sub-system technologies are evolving to meet the stringent requirements. Some such examples are high accuracy star trackers for attitude pointing and control, large detector arrays, microstrip antennae, high-speed microprocessors, high-efficiency propulsion systems and laser communications for high data throughput. To meet the future large demand for Nanosatellites, existing industries and new entrants are gearing up for mass production of components and sub-systems both in India and abroad.

This emerging 'new space' scenario is innovative and is poised for a major change in the application arena catering to a large population in the world including the unreached, underdeveloped areas of Africa, Asia, South America, Arctic and Antarctic regions.

The Nano satellites have adequately demonstrated their capability to adapt to diverse applications in the areas of Communication, Science, Earth observation, Education and Health care services at a much lower cost. This will be the order of future with the greater involvement of developing countries with students and academia, government agencies and new industries/companies, for a sustained universal access to Space technology and applications, as envisioned by the founding father of Indian Space Program, Dr. Vikram Sarabai.

REFERENCES

1. Modern Small Satellites-changing economics of space: Prof. Martin Sweeting, Surrey Space Centre, UK.
2. Nanosatellies-The Tool for a New Economy of Space: Opening Space Frontiers to Wider Audience, Journal of Aeronautics and Aerospace Engineering.
3. NASA-GSFC Nano-satellite technology for earth science mission, Acta Astronautica

* * *

L.S. Satyamurthy has worked as a satellite quality system expert since the beginning the Indian satellite program. He has extensive experience in international technical liaison, space technology applications in health care and commercialisation of space products and services. He has served as Scientific Diplomat at the Embassy of India in USA, Director, Business Development at Antrix Corporation and as Program Director, Telemedicine, ISRO.

CHAPTER 02

MISSION PLANNING, ANALYSIS AND DESIGN

T.K. Sundaramurthy, H.N. Bhagavan & S. Rangarajan

Any satellite, big or small, holistically comprises two important domains- a 'Project' and a 'Mission' - which are closely knit functional areas. The project encompasses satellite design, development, manufacture and launch. Whereas the mission covers all activities from the definition of goals/objectives, interlink between satellite, launch vehicle and ground segments culminating in the successful operation of a satellite in orbit. It is important to have a clear understanding of satellite orbits for mission planning and analyses. This will help in arriving at optimal specifications for a Nanosatellite system vis-a-vis constraints of weight, volume and power and is a challenging exercise in system engineering.

Conventionally a mission team is one entity in an agency which functions and interacts closely during the satellite system design, development, manufacture, launch and orbital phase, till the satellite is declared operational. Subsequently, the identified mission and network control centre and the respective service providers carry the baton for the effective delivery of a satellite's service to the end-user/ customer. Mission operations is a multi-disciplinary science covering satellite orbital mechanics, launch vehicle dynamics, ground station and communication engineering.

Readers will definitely experience the satellite orbital journey in this chapter.

2.1 OVERVIEW

The success of any mission depends on how well it is planned, analysed and executed. It is all the more so for a satellite mission, which is a complex and multi-disciplinary activity. The mission life-cycle starts right from its conception and goes on until its last use. The activities may broadly include:

- Definition of mission objectives
- Preliminary estimate of mission needs, requirements and constraints
- Identifying mission architectures, including alternatives
- Identifying system drivers
- Mission analysis

- Flight dynamics and launch window analysis
- Generating flight operations and contingency recovery procedures
- Simulation and operators training
- Real-time operations and satellite health monitoring
- Mission utilisation up to the end of the mission's life

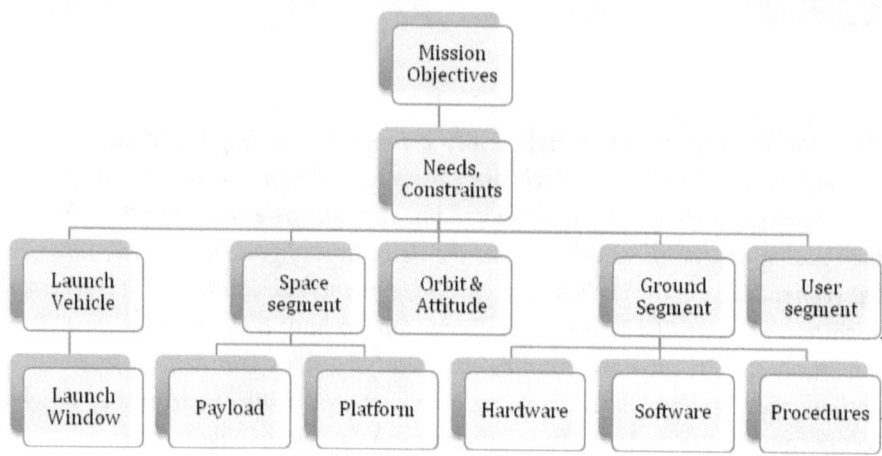

▲ **Fig. 2.1:** Mission elements of any satellite

Fig. 2.1 shown above is valid for any satellite mission. For a Nanosatellite mission, we need to scale down from the well-established procedures for large satellite missions to what is appropriate for the smaller one. For example, some of the elements like the launch window that would be a requirement for a dedicated mission will be handled as a constraint in a Nanosatellite mission. If the Nanosatellite is just a piggyback on the launch vehicle mission, the primary payload would dictate the orbit in which the Nanosatellite will be injected. Within the constraints of weight, volume and power, arriving at the optimal specifications of each satellite sub-system to meet the mission goals is a challenging exercise in system engineering.

The ground segment consists of earth stations and the satellite control centre. The satellite operations' control centre will monitor the satellite's health, send commands to the satellite as needed and collect tracking data for orbit information. The user segment includes the ground station, payload data reception/processing/archival and interface for user interaction. In a Nanosatellite mission, to lower the post-launch operational cost and complexity, it is desirable to eliminate some or combine several of these sub-segments into one.

2.2 BASIC CONCEPTS OF FLIGHT DYNAMICS

Whether it is big or small, the basic definition of a satellite is that it should be going around the earth in a closed path, called its orbit. The moon, for example, is a natural satellite of the earth and it goes around the earth at a distance of about 375,000 kms and it takes about a month to go around once and this time duration is called its orbital period. It is very important to note that for going round the earth, the moon does not consume any fuel and the motion happens due to gravitational forces and the initial conditions. The same thing holds good for any man-made satellite as well.

Satellites, orbiting in circular paths around the earth, have periods increasing as they go away from the earth as illustrated in the table below:

Altitude (km)	400	800	20,200	36,000	375,000
Period	92.5 min	100.7 min	12 hour	1 day	~28 days
Type	LEO	LEO	GPS/MEO	GEO	Moon

The velocity vector constantly changes its direction due to the centripetal force of the gravitational pull directed towards the centre of the earth. In a circular orbit, this constant change in the direction of the velocity vector matches the curvature of the earth. The curvature is more in a LEO compared to a more distant orbit like GEO and the speed is higher for orbits closer to the earth, e.g., the near-earth satellites move at speeds of around 7 km/sec, whereas for a geostationary satellite it is around 3 km/sec.

2.2.1 Orbital Elements

All satellites orbiting the earth move under the same equation of motion derived by equating the acceleration to the gravitational pull from the centre of the earth divided by its mass. This second-order differential equation can be solved for the orbital motion if the initial position vector and initial velocity vector (which together are called the state vector) are given. For example, this could be the satellite injection condition of the launch vehicle (i.e. injection point and the injection velocity). As these two vectors are both in three-dimensions, a total of six numbers distinguish one orbit from another. These six numbers are referred to as 'orbital elements' or 'orbital parameters' as explained below.

Before going into these orbital parameters, there are two important references to define some of the orbital parameters, namely - vernal equinox and line of nodes as shown in **Fig. 2.2**.

- Vernal equinox is the vector in the celestial sphere from the earth's centre towards the sun when it crosses the earth's equator while moving from south to north which occurs on March 21/22. The celestial reference frame used for orbital mechanics is defined by the coordinate system where the vernal equinox is the X-axis, true north is the Y-axis and third orthogonal is the Z-axis.
- The portion of the satellite orbit where it moves from the North Pole to the South Pole is referred to as 'descending node' and the other portion where it moves from the South Pole to the North Pole is referred to as 'ascending node'. Each node crosses the earth's equatorial plane at one point and the line joining these two points is called the line of nodes.

A satellite orbit can be described by the following six parameters:
- Semi-major axis - **a**
- Eccentricity - **e**
- Inclination – **i**
- Right ascension of ascending node - Ω
- True anomaly - ϕ
- Argument of perigee - ω

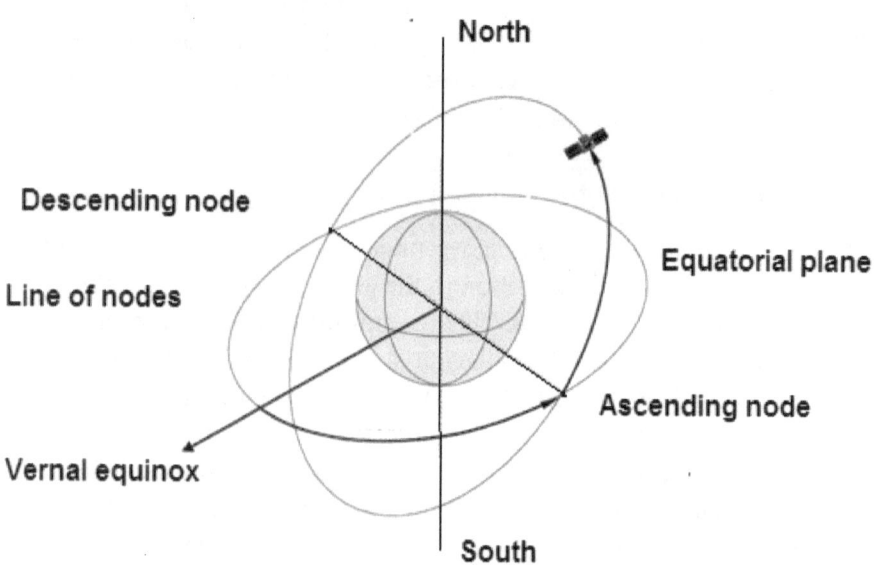

▲ **Fig. 2.2:** Vernal equinox and line of nodes

The semi-major axis gives the distance of the satellite from the centre of the earth (altitude is the distance of the satellite from the surface of the earth). The perigee is the nearest point and the apogee is the farthest point of the satellite from the earth. At the perigee, the satellite is moving at its fastest and at the apogee at its slowest velocity.

The shape of the orbit is defined by Eccentricity 'e' which can vary between 0 to 1. Most of the satellites (including Nanosatellites) placed in LEO and GEO orbits are circular where e = 0.

Inclination 'i' is defined as the included angle between the orbital plane and the equatorial plane. The value of 'i' is in the range of 0-180 degrees. Generally, if the value is lesser than 90 degrees, the orbit is called prograde and if the value is greater than 90 degrees, then it is called retrograde.

The right ascension of the ascending node (RAAN) Ω is defined as the angle between the vernal equinox and the point of the ascending node along the equator.

The true anomaly ϕ (Phi) is defined as the angle subtended between the perigee and the satellite's location in the orbit.

The argument of perigee ω is the angle between the ascending node and the perigee. If both the perigee and ascending node are at the same point, then the argument of perigee will be zero degrees.

These orbital parameters are pictorially shown in **Figures 2.3(a) and 2.3(b)**

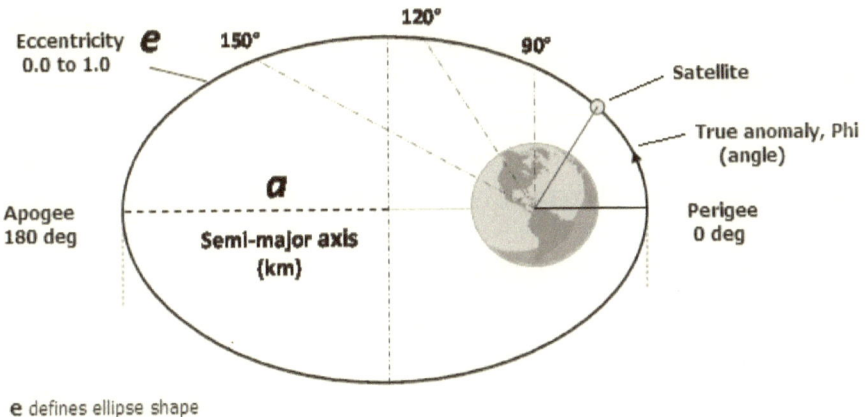

e defines ellipse shape
a defines ellipse size
Phi defines satellite angle from perigee

▲ **Fig. 2.3(a):** Orbit parameters

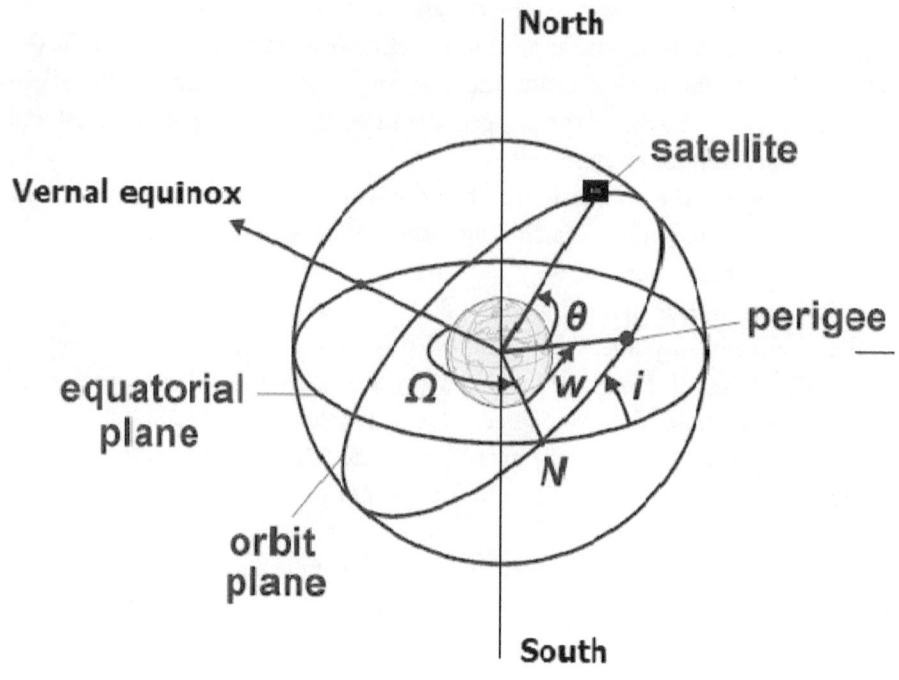

▲ **Fig. 2.3(b):** Orbit Parameters

2.2.2 Types of orbits

Satellites, based on their mission objectives, are placed in different types of orbits, which are broadly classified as (i) Low earth orbit (LEO); (ii) Medium earth orbit (MEO) and (iii) Geostationary orbit (GEO). Apart from this, planetary missions are launched into specific orbits around the corresponding planets (Venus or Mars). Various types of satellite orbits are shown in **Fig. 2.4.**

2.2.2.1 Low earth orbit (LEO)

Satellites in low earth orbits circle the earth at a height of 500 to 1,000 kms above the surface of the earth. The inclination, which is the angle between the orbit plane and the equatorial plane, may vary between 0 to 180 degrees. If the orbit is at an inclination of about 90 degrees, it is called a polar orbit and will have worldwide coverage. The period of such an orbit will be in the range of 90 to 120 minutes depending on the exact orbital height. In inclinations between 0 to 90 degrees, the satellite travels in the same direction as that of the earth's rotation and is termed as prograde orbit. If the inclination is greater than 90 degrees, the satellite travels opposite to the earth's rotation and it is called a retrograde orbit. Most of the earth observation remote sensing satellites are placed in LEO near-polar orbits to provide global coverage.

Sun-synchronous orbit (SSO)

By proper selection of altitude and inclination, an orbit can be achieved where the orbit precession rate is matched with earth's rate around the sun (0.9856 deg/day). This orbit is called the sun-synchronous orbit, which is a special case of LEO in which the orbit plane maintains a constant angle with respect to the sun's vector. This orbit is very useful for remote sensing/earth observation satellites with optical sensors to monitor the earth's natural resources. This orbit ensures:

a) The satellite passes over a given location on earth at the same local solar time guaranteeing the same illuminating conditions varying only with season.

b) The satellite in this orbit covers the whole surface of the earth being polar in nature.

2.2.2.2   Medium earth orbit (MEO)

Medium earth orbit satellites are at a distance of approximately 1,000 to 20,000 kms above the surface of the earth. The period of such satellites is in the range of six to 12 hours. The Global Positioning System (GPS) satellites of the USA are launched into medium earth orbits at an altitude of 20,200 kms and they make two revolutions per day around the earth. MEO orbits have also been used for communications by Iridium company, the USA by placing 66 satellites with proper separation.

2.2.2.3   Geostationary earth orbit (GEO)

A geosynchronous orbit is a prograde orbit with the orbital period same as that of the earth, that is, 23 hours and 56 minutes. A special case of GEO is with the plane of the orbit being very near to the earth's equatorial plane, where the satellite will appear to hover around the same point with respect to the earth. Such an orbit is very useful for satellite communications. The orbital height from the surface of the earth is 35,786 kms and the period of revolution is 23 hours 56 min. The GEO orbit is extensively used by all countries for national and international communications, including telephones and direct to home TV broadcasts. From this altitude, the earth's disc subtends an angle of about 17 degrees and various types of sensors are flown on satellites used for meteorological and weather monitoring.

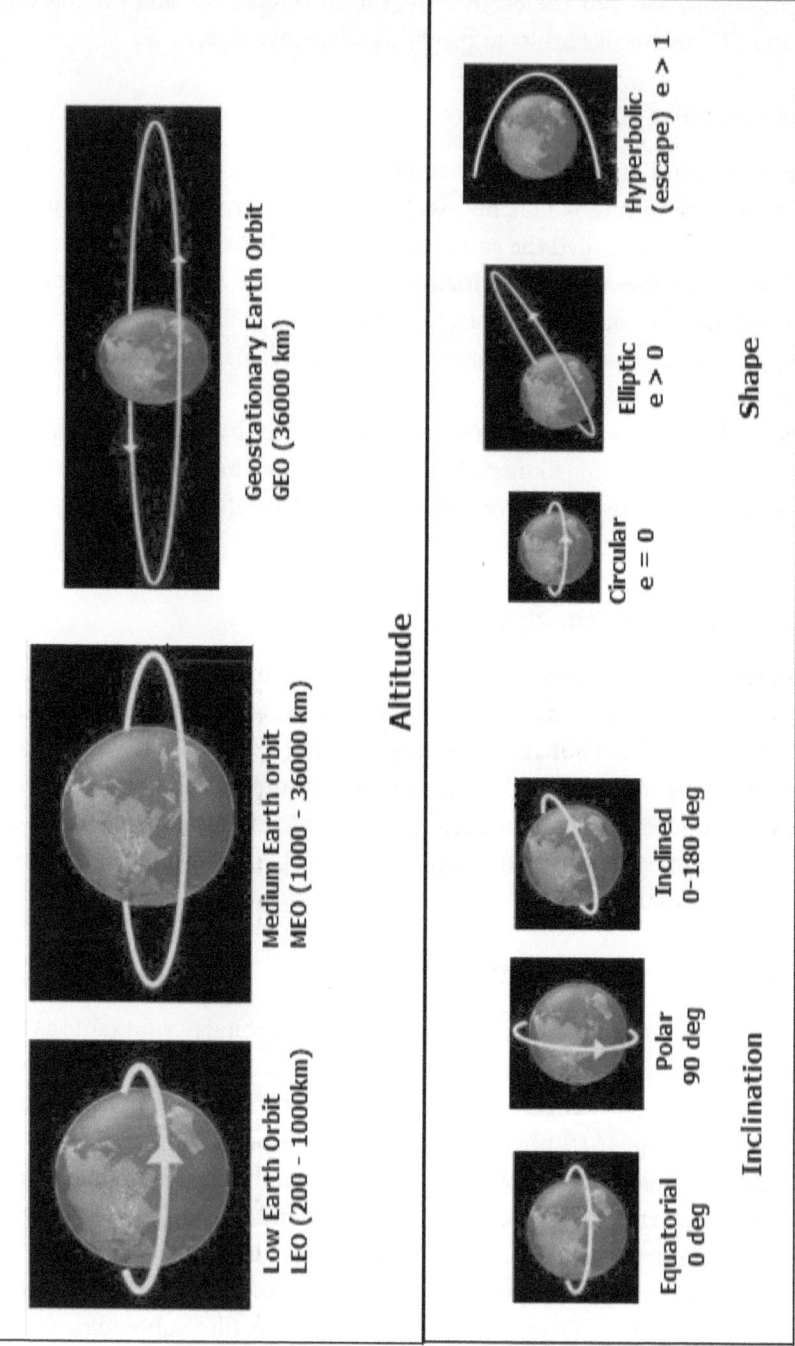

▲ **Fig. 2.4:** Types of satellite orbits

2.2.3 Satellite observation geometry

When a satellite is tracked from the earth, the apparent motion is the combination of the satellite's motion in the orbit as explained in the previous section as well as the earth's rotation on its own axis, which has a period of one day. At a specific altitude over the equator, we obtain a condition of apparent stationarity of a satellite as observed from the earth, when the earth's rotation on its own axis matches with the orbital motion of the satellite. In this unique case of the geostationary orbit (GEO) these two effects cancel out. For example, a DTH operator installs a dish antenna on the roof of a house to receive TV signals from a GEO satellite and the antenna is aligned to a fixed direction of the satellite signal, which does not move at all. Other than GEO satellites, all the other satellites have to be tracked with an ever-changing direction of the antenna towards the satellite.

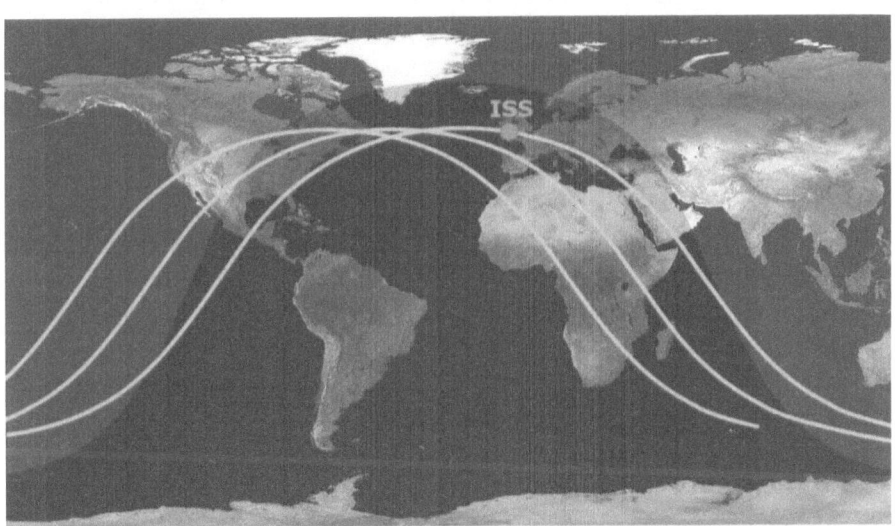

▲ **Fig. 2.5 Satellite observation geometry (Courtesy:** Science ABC)

The point that is vertically below the satellite is called the sub-satellite point (SSP) or nadir. As SSP is a point on the earth's surface, it can be uniquely specified by its latitude and longitude coordinates. A satellite's ground trace is the path over the earth's surface, obtained by joining the SSP's in successive instants. If the earth is not rotating, the ground trace would have been a great circle on the earth's surface. As the satellite goes all the way around the earth during each orbit and the earth also rotates on its own axis, the ground trace is not easy to imagine. The earth's rotation constantly shifts the SSP towards the west and over an orbital period, the ground trace will not be a closed path. For a LEO satellite of 100 minutes period, the equatorial shift per orbit is around 25 degrees westward at the equator as shown in **Fig. 2.5.**

The figure gives the ground trace of the International Space Station (ISS) over three orbits. The ISS is at an altitude of about 370 kms and inclination of 51.6 degrees. The ascending node shifts westward about 25 degrees, the value of the earth's rotation in the one orbital period. It is noted that the maximum latitude of the SSP is the inclination of the orbit as seen in the figure.

For an ideal GEO orbit, the eccentricity is 0 since it is a circular orbit and the inclination is 0 as the orbital plane lies in the equatorial plane. Hence its SSP is a fixed point on earth's equator, i.e. with latitude = 0 and longitude = φ. Just this one parameter φ, distinguishes one GEO satellite from another. If the inclination is not exactly zero, but the period is synchronised with the earth rotation, we get a geosynchronous satellite (GSO), whose ground trace is a figure of eight.

2.2.4 Satellite look angles

The direction wherefrom a satellite signal is received at any given earth station is identified by two angles, namely elevation and azimuth. The elevation angle specifies how far up in the sky the satellite is with respect to the local horizon and can range from 0 to 90 degrees. An elevation of 0 degrees is towards the horizon, whereas an elevation of 90 degrees is vertically overhead. The azimuth angle is measured from the local north direction towards the east and has a range of 0 to 360 degrees. Thus, the due east direction will have an azimuth of 90 degrees and the due South 180 degrees. Taken together, azimuth and elevation angles with respect to the given earth station specify the direction towards the satellite at any given instant. Obviously, if the elevation at a given time is negative, the satellite is invisible to that station at that time instant as it is below the horizon.

For an ideal GEO, the azimuth and elevation angles do not vary with time. On the other hand, for a LEO satellite in a near-polar orbit, there would be four to five slots of 10-15 minutes over a day in which the satellite would be above the horizon from an earth station such as Bengaluru. Each such visibility stretch is called a "pass". During a pass, the satellite earth station can receive the telemetry and payload data streams as well as be able to send any telecommand to the satellite if needed. If more real-time contact with the satellite is desired, it will require more tracking stations added across the globe and this will be an expensive proposition for a Nanosatellite project.

Figure 2.6 shows a typical pattern of visibility over a day for one of the Iridium satellites for a near-equatorial earth station. The orbital height is 781 kms, eccentricity is 0, and inclination is 86.4 degrees. The total radio visibility of this satellite over this station is only about 45 minutes per day, split into typically two passes in the ascending part of the orbit and two in the descending, which occur roughly 12 hours apart.

But the Iridium constellation has 66 satellites in six orbital planes and so the same ground station will have more than one satellite of the constellation over its horizon at any given time. In a LEO constellation, satellites maintain constant positions relative to each other.

Such constellations are mushrooming these days, e.g., Starlink of SpaceX, the USA alone has filed for 41,943 satellites for providing internet services globally and launches 60 satellites every other month!

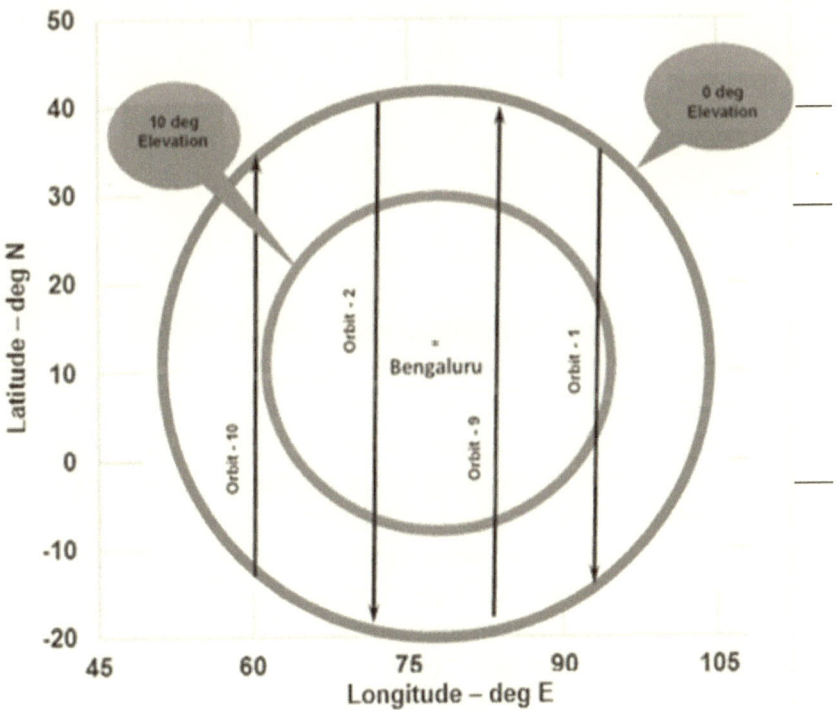

▲ **Fig. 2.6:** Typical radio visibility of a satellite from the ground station

2.3   MISSION DEFINITION AND MISSION OBJECTIVES

A mission design involves defining mission objectives, developing a mission concept, designing orbits and trajectories and evaluating the space environment of the satellite's performance. A well-articulated mission definition leads to the identification of specific primary and secondary mission objectives. Mission objectives identify what the ground crew, spacecraft and payload must do for mission success. The figure below depicts the major steps in a mission design:

Identify & prioritize user needs

Define constraints

System characteristics

Functional requirements

Technical specifications

System Architecture

Decompose requirements to lower levels

Iterate and refine

In the case of a Nanosatellite, it could be a specific technology demonstration, or imaging target areas on the surface of the earth for a specific analysis or a study of the earth's environment, space science investigation or simply to gain the experience of building and operating a satellite mission. In the early years of Nanosatellites, it was driven more by the excitement to build and launch one's own satellite and the design was often ad hoc. As technologies matured and the number of such missions increased, more system engineering concepts are being applied, better performing COTS are used, constraints like inability to de-orbit are taken into account and better missions have been accomplished. They have advanced from being perceived as only suitable for educational purposes to robust platforms for conducting space missions with a good cost to benefit ratio. They have been used for a variety of applications in addition to contributing to academia. Small satellites have also served as testbeds for numerous innovative concepts and mitigating risks in operational missions.

In the case of Nanosatellites, the payload should preferably be a single sensor instrument, mature technology, simple to calibrate, instrument dedicated resources of power, pointing and data storage, only a few payload modes, simple pointing and low rate data output.

2.3.1   Mission definition of components, interfaces and tasks

Nanosatellite missions increasingly use system-engineering concepts. The top-level, requirements and operational scenarios lead to the high-level functional diagram. At this stage, one could sketch the N^2 (N-squared) diagram, which allows traceability from requirements to specifications and vice versa. The process of identifying non-conflicting components and their corresponding interfaces through the N^2 diagram leads into developing a mission over-head model in terms of power, telemetry and computation. A typical N^2 diagram looks like the one shown in **Fig. 2.7.**

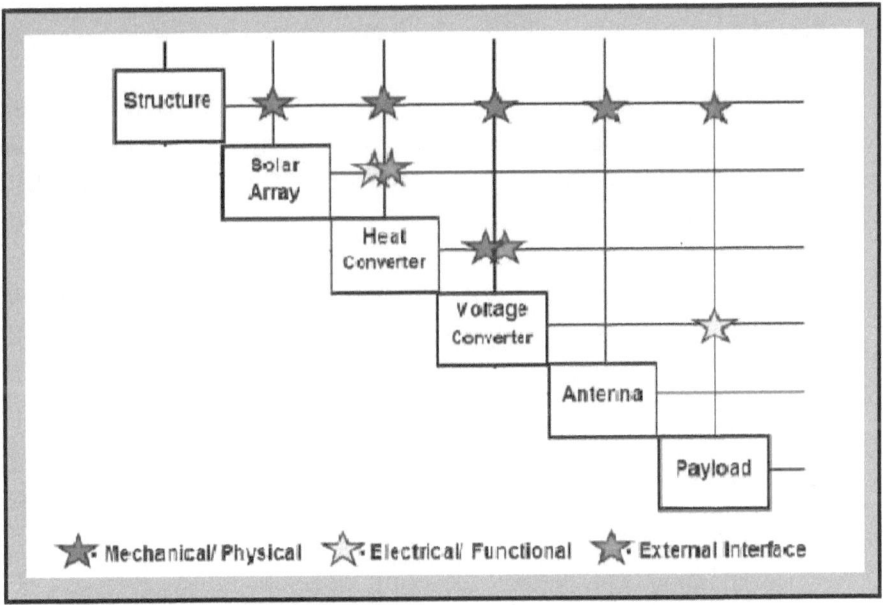

▲ **Fig. 2.7:** Typical N2 diagram for a Satellite mission

The N^2 diagram is a matrix representation of interfaces between the elements of the system. System elements are on the diagonal with inputs represented as vertical lines and outputs as horizontal lines. This sort of diagram can quickly pinpoint where conflicts could arise in interfaces. Interfaces can support one-to-one, one-to-many or many-to-many type connections. They can be electrical, which includes power and communication interfaces, or mechanical. This approach identifies limitations early in the design process so that the correct compromises can be made. For example, if an instrument's aperture has a field of view of 20 arc seconds and the spacecraft has the pointing capability of 30 arc seconds, one can infer that the operators can never be sure if the object is in the instrument's field of view.

2.3.2 Principal system drivers

Once the mission objectives are finalised and approved by the concerned authorities, the workflow proceeds as follows:

- Based on the mission objectives, one derives the mission requirements, both functional and performance as well as the constraints, which could include external drivers like cost, schedule and prior lessons learnt.
- These then translate as inputs for the space segment and ground segment configurations.
- Generation of "orbit and attitude accuracies requirements" based on mission needs. This, in turn, forms inputs for configuring ground tracking systems, sensing systems, onboard attitude control and their choices. Standard models for the environment are used for orbit propagation and disturbance torque estimation.
- Identification of ground network including TTC network, data reception, processing and communication systems.

Program constraints in Nanosatellites could include:
- A specific tracking network may have to be used
- Existing flight hardware to be used
- Existing ground systems to be used and spacecraft design to be made compatible with them
- Educators and the academic community, including students, have to be involved

The table below illustrates how the different sub-systems, such as structures, thermal design, sensors and attitude control, communication links or orbital geometry may be impacted by the given mission objective.

Mission Objective	Principal System Driver
Systematic repetition of satellite imagery for temporal analysis	Orbit and payload swath
Space Astronomy; observe external stellar targets	Structure to support micro electro-mechanical systems (MEMS) based deformable mirror in space is a promising technology to enable the extreme wavefront control required for direct imaging.
Store and forward message delivery	Onboard antenna and communication system
Optical communication for ~100 Mbps link to ground	Pointing accuracy; tracking & stabilisation. Atmospheric turbulence
Collection of high-resolution visual sensor data	Thermal design of focal plane components
Complex operations to be performed by the OBC	Increased robustness to radiation
Greenhouse gas monitoring	Onboard storage and intelligent dumping
Internet of Space Things and Rural connectivity	To support software-defined networking and network function virtualisation

2.4 MISSION ANALYSIS

Mission analysis are carried out covering a wide range of aspects and the respective sub-system teams are required to participate in these deliberations. For example:

- Sun aspect angle variation with seasons, sunlit-eclipse duration of orbit, payload operational modes, satellite orientation and shadow analysis will provide inputs for thermal, power, communication and AOCS designers.
- Sun, moon and earth shadow occultation into the field-of-view will provide inputs to sensors' designers.
- Satellite elevation angles for the ground station and duration of radio visibility for each pass, number of passes over a ground station will provide inputs for TTC designers and ground system operators.

The best way to understand this trade-off is to consider a few case studies.

2.4.1 A Remote sensing mission

Many remote sensing platforms are placed in an orbit called sun-synchronous such that they cover each area of the world at a constant local time of the day so that images taken on different days have similar solar geometry for ease of monitoring temporal changes or for mosaicking adjacent images. A sun-synchronous orbit requires an inclination close to 100 degrees and hence the orbit is near polar.

As the satellite goes around the earth, the remote sensing imager "observes" a certain part of the earth's surface, referred to as the swath. Imaging swaths could be tens or hundreds of kilometres wide. As the satellite moves in its orbit, this swath push-brooms along the ground trace. The earth's rotation on its axis automatically moves the swath to cover a new area in consecutive orbits. The orbital height decides the orbital period and hence the longitude shift per orbit. For a given swath of the imagery, one can then work out the orbital cycle time, namely the number of days after which the same ground area will be imaged next.

In near-polar orbits, areas at high latitudes will be imaged more frequently than the equatorial zone due to the increasing overlap in adjacent swaths as the ground traces come closer together near the poles. The orbital cycle is an important consideration for a number of monitoring applications, especially when frequent imaging of the same area is required. Using a steerable imager, a satellite-borne instrument can view an area (off-nadir) as well, thus making the 'revisit' time less than the orbit cycle time.

If, for example, a Nanosatellite has the mission objective of imaging for monitoring a specific lake for pollution in different seasons, it should look for piggybacking on an operational remote sensing mission with acceptable orbital cycle time and local time of imaging. The amount of data collection over the region of interest can then be increased if the payload can have a steering capability along track and/or across track. This secondary requirement would, in turn, impact the attitude control system, which could affect the power generation and so on.

2.4.2 Space Astronomy Mission

As an example consider FlatOSat (Flat Optics CubeSat) mission of John Hopkins University. The overall mission objective was to build a Nanosatellite with far-field imaging and to do better than the diffraction-limited imaging. From the mission objective, one derives target values for key performance parameters like angular resolution, angular sensitivity, orbit average power and attitude determination accuracy.

The mission analyses led to several trade-off studies for this mission and are summarised below:
- Identified 33 requirements
- Identified four alternatives
- Formulated five selection criteria
- Used pair-wise comparison to establish criteria weights

- Prepared utility functions
- Evaluated alternatives
- Performed sensitivity analysis
- Eliminated sensitivities
- Selected preferred alternative

2.4.3   INS- 1C mission

INS-1C was an ISRO initiative to design and develop a modular Nanosatellite bus for technology demonstration as well as to provide an opportunity for building innovative payloads for universities with a quicker turnaround. INS-1C carried a miniature multispectral technology demonstration (MMX-TD) payload with imaging in RGB (Red, Green and Blue) bands with highly compact multi-fold optics. The payload data was downloaded @1Mbps through an S-band carrier. The satellite had two deployable solar panels in the positive and negative roll faces, along with a body-mounted panel and these together could generate 27 Watts at normal incidence. The sun pointing was the nominal attitude with the negative pitch towards the sun and the positive roll along the nadir x sun due to power constraint. In the sun pointing orientation, the angle between body yaw and the earth could be as high as 100 degrees whereas it would be 65 degrees in the sun-tracking attitude with the positive yaw along the nadir and positive roll along the nadir x sun. As the S-Band antenna beam width was 60 degrees, the sun tracking orientation was the preferred attitude for S-Band data download. Further, the amount of rotation involved between the sun pointing and the sun tracking orientation was less compared to the sun pointing and earth pointing orientation. A typical example of mission analysis that was performed to arrive at the optimal bus design under conflicting considerations.

2.4.4   Scaling down missions from large to small

A Nanosatellite design exercise can come to fruition only by tailoring requirements to realistically fit into the limited resource envelope. A pressing challenge faced by Nanosatellite communications is the limited bandwidth that leads to low data rates, high latency and eventual performance degradation. Hence AI techniques and deep learning are being studied for their usefulness in designing a Nanosatellite.

For attitude stabilisation, one option - no stabilisation hardware at all - is a simple and therefore an attractive alternative for small-satellite applications. However, this choice will lead to pressing challenges in other sub-systems. To achieve downlink margin and adequate power in any attitude, the satellite's antennae must have omni coverage and solar cells must be distributed on the entire surface of the satellite.

The ground control for a small satellite can be simple enough to be built around the user's personal computer and operated as a computer-controlled laboratory apparatus. Sophisticated ground-station software eliminates most of the need for operator interaction and provides a simple, user-friendly interface requiring little special training. The computer is the only interface to the ground-station equipment, so the user need not understand any other equipment interfaces.

2.5 MISSION OPERATIONS

Mission operations are activities performed post launch by the operations teams, including an integrated system of hardware, software, people and procedures that must act in unison to accomplish the mission goals. Though mission operations focus on the period after launch, substantial work must be done in all phases of the mission to prepare for the operations. It is important to involve the mission personnel right from the early design stages, as they can give a different perspective of looking at what can go wrong, whereas designers generally expect things to work!

Mission operations planning encompasses a wide range of activities as shown below:

Function	Activity
Mission planning	Decide what to do and when
Activity planning	Different scenarios, sequence of events, FOP, CRP
Mission control	Managing daily operations
Flight dynamics	Determine where the satellite is at any given time for planning operations
Command operations	Managing satellite bus and payload
Data services	Communication links, data flow, information extraction, data archiving

In fact, planning a mission operation is one of the most important deliverables before launch. Timely definition of payload characteristics can be achieved by including payload designers and mission planners while developing the mission concept. The mission planning phase concludes with the finalisation of the mission readiness review (MRR) after which a mission team is declared as being ready for carrying out mission operations on the satellite.

2.5.1 Pre-launch phase operations

The pre-launch phase starts when a satellite successfully undergoes all environmental tests and is shipped to the launch site. At this stage, the mission team has to ensure that all ground systems have undergone tests as well as evaluation and are ready for operations. All databases of telemetry channels, telecommand codes and sub-system power consumption details are updated in the mission documents based on the actual test results of the satellite. In case of any deviation from the normal performance or any other contingency situation affecting the health of the satellite, an in-depth analysis even during the pre-launch phase is carried out to identify the cause so as to take up appropriate corrective actions quickly in anticipation of such anomalies that may occur post launch. At the launch base, the satellite undergoes limited electrical tests before integrating with the launch vehicle. In the case of satellites that are to be kept in the power OFF mode during the launch phase as is the case for most Nanosatellites, the uplink and downlink interfaces are tested from the launch pad to ground station as part of the launch countdown.

The major pre-launch phase activities include:

- Generation of satellite *simulation data* as well as to conduct pre-launch simulation and testing of ground network hardware and software as well as training of operational personnel.
- Generation of *operational procedures* for all bus systems and payloads.
- Generation of *quick look display schemes* for both real-time and (near real-time) NRT health monitor and analysis.
- Generation *of operational procedures* including contingency requirements.
- Generation of *detailed flight sequence for initial phase* operations.
- Generation of *detailed communication link operations* and formats for information exchange.
- Generation of *routine operations plan,* including data collection and data dissemination.
- Generation of *overall mission management procedures.*
- Plan for health analysis, archival and dissemination of satellite data.

2.5.2 Initial phase operations

Initial phase operations start immediately after injection of the satellite into the orbit by a launch vehicle. In most of the cases, a Nanosatellite is powered ON at this stage automatically after receiving the 'snap' signal from the launch vehicle. With satellite power ON, all onboard systems get actuated except the payload. Based on the injection parameters provided by the launch vehicle team, a preliminary orbit determination (POD) is carried out. At this stage, the satellite will usually be 'tumbling' in space with high body rates and it has to be stabilised using actuators like magnetic torquers by automatically selecting the 'de-tumbling' mode of the attitude control system. Once the attitude of the satellite is stabilised and it comes into the radio visibility of the ground station, both the downlink (telemetry) and uplink (telecommand) signals are established and the planned operations are carried out. Orbit determination of the satellite is an important activity for which either the tracking data from different ground stations is used or the 'two-line element' (TLE) data from NORAD (North American Aerospace Defence Command) is used. After verifying the performance of all sub-systems, including the payload and its data quality, the operational phase starts.

2.5.3 Normal operational phase

The operational phase continues until the end of the mission and/or satellite is decommissioned. During this phase, regular monitoring of health parameters of the satellite, operating the payload as per plan and user requirements, maintaining the orbit and handling any contingencies are carried out. In case any updating of onboard software parameters (such as algorithm coefficients, payload gain or software patches) is required, the same is converted into data commands and uplinked to the satellite. It is important to keep track of some critical parameters (such as bus voltage and current or sub-system temperatures) and automatic alarms are set to alert an operator in case they exceed set limits. It is also important to record any changes made in the configuration of the satellite.

2.6 MISSION CONTROL SOFTWARE

2.6.1 Satellite health monitoring and control

This set of programmes address the need to acquire telemetry data, validate, process, monitor, analyse and control the satellite's health in real-time and near real-time. The health will be presented in a variety of ways, including graphics or tabular trends. It will also support real-time command and verification. It should also include modules for data archival and retrieval. It should interface with the flight dynamics modules for ground segment support. Specific attention should be given to the following:

- Automation plan for operation of sequence of events (SOE), flight control procedure (FCP) and contingency recovery procedure (CRP).
- Plan for user-friendly payload programming systems.

- Plan for the assessment of satellite behaviour and generate timely alerts to the concerned

Planners predict trends and performance, trying to forecast faults before they might occur and then plan preventative action. This task can involve a software simulator with elaborate models for telemetry, power, thermal, dynamics and other sub-systems.

We must trade-off automation against ground crew operations in the ground segment as well as the space segment. Using software in place of people does not always reduce costs but we need to assess this option. It is important to provide a "safe mode" to conserve power when unexpected events occur during the non-visible portions of the orbit.

2.6.2 Flight dynamics software

Based on the ground tracking data or onboard satellite position system, a satellite's orbit is determined and updated periodically. From the determined orbit, it is required to generate the ephemeris, orbital events such as visibilities for ground stations and eclipse details. If the satellite has star sensors, one could compute the attitude using the data. Most Nanosatellites have software for onboard orbit computation using numerical integration.

The flight dynamics software system can be broadly categorised as:
- Orbit-related
- Attitude-related
- Onboard software support

In each of these areas, the software support has overall functions such as:
- Regular operations
- Orbit and Attitude information
- Parameters for uplinking
- Information generation that helps evaluation of ground and onboard orbital software performance

* * *

T.K. Sundara Murthy has vast experience in mission planning, analysis and execution of communication and meteorological satellites of ISRO. His specialisation includes mission analysis, satellite orbit and attitude determination and control, satellite propulsion and thermal analysis.

H.N. Bhagavan is an expert in handling ground station network, satellite data management and satellite control centre operations for various communication satellites. His area of expertise is in satellite mission management including mission operations during pre- launch and in-orbit phase covering satellite health monitoring and analysis.

Dr. S. Rangarajan, a renowned nuclear physicist from Tata Institute of Fundamental Research (TIFR) is a satellite technology veteran with four decades of experience in the fields of satellite communication, Mission Management, mission planning and analysis as well as satellite and launch vehicle ground station management. In his illustrious career at ISRO, he has held several important executive positions. He has also served as Senior Vice President, World Space, the USA for satellite digital radio broadcasting.

PAYLOAD OPTIONS FOR A NANOSATELLITE

M. Venkata Rao

Every satellite mission conceived and executed by any agency, industry or academia will have a specific purpose - whether it is for research, technology demonstration or specific business services. While a satellite is one monolith, the main engine and the driving force is the payload.

The payload is a multifunctional instrument system which can observe, assimilate, process and deliver service to the end-user directly or indirectly through the ground system network. The payloads for the Nanosatellite are generally categorized for remote sensing application like a camera or for communication service like a transmitter-receiver combination termed as a transponder or for scientific exploration like an instrument sensor. In the case of planetary or interplanetary missions, the payload is a combination of all these systems. Majority of the Nanosatellites launched so far have carried remote sensing payload such as an optical or infrared camera for earth imaging applications.

Payload engineering is again a multi-disciplinary field of science and technology with specific emphasis on the aspects of physical engineering such as optoelectronics/instrumentation technology, communication/microwave engineering and astrophysics.

Readers will get good information on the payload technology in this chapter.

Any space mission is launched with a certain intended purpose, for example, a rocket is launched to place a satellite in a specified orbit and a satellite is launched to provide communication service, atmospheric studies or to make earth observations. This purpose, also known as 'mission objective', is met with one or more instruments called **payload**. Therefore, the definition of mission objective(s) and identification of the payload for any satellite project in its 'inception stage' is very important. In case of Nanosatellites, though there are limitations in terms of size, weight and power (SWaP), various payloads flown in such satellites world over have demonstrated that several new concepts and technologies can be proven, which may be later adopted in larger operational satellite missions. This is possible due to new technology developments such as miniaturisation of electronic components including high-speed/high-density VLSI devices, lightweight materials, miniaturised optics as well as sensor devices and micro electro-mechanical devices (MEMs).

Selection of payload for a Nanosatellite project, particularly by an academic institution, depends on several factors, besides meeting the mission objectives. These factors include technology involved, expertise/resources available within the institution, complexity of ground data reception, processing and analysis/interpretation, cost and schedule. It is also important to note that the configuration of Nanosatellite bus elements such as structure, power, communications and attitude determination and control system (ADCS) will largely depend on the selected payload configuration. A few payload options for a Nanosatellite project including their basic principles, applications and other technical elements are covered in this chapter.

3.1 TYPES OF PAYLOADS FOR A NANOSATELLITE

Payloads for a Nanosatellite can be broadly classified into the following categories:
- Earth observation (EO) payloads
- Communication payloads
- Scientific payloads

Majority of Nanosatellites developed so far by various academic institutions, both in India and abroad, have carried earth observation payloads such as an optical or infrared camera for imaging earth features as can be seen from the list of student satellites launched by ISRO given in **Table. 3.1**. Other EO payloads include spectrometers for pollution monitoring, microwave radiometers and radio occultation sounders for atmospheric studies.

A few examples of communication payloads include point-to-point message store and forward systems, characterisation of ionospheric effects on RF signals and identification of ships through their radio signals. Recent trends show some interest in laser communication payloads but mostly in a one-way direction (space to earth).

Scientific payloads include biological experiments under near zero-gravity conditions, observation of specific stellar objects and study of natural disasters such as earthquakes or volcanoes.

▼ **Table. 3.1:** Nanosatellites launched by Indian academic institutions:

SI No.	Institution/ University	Name of Nanosatellite	Payload flown
1	Nurul Islam University	NIUSAT	3-band (RGB) camera
2	IIT, Bombay	PRATHAM	Measurement of TEC in the ionosphere
3	PES University	PISAT	CMOS Camera
4	Satyabhama University	SATYBHAMASAT	Infra-red spectrometer
5	College of Engg., Pune	SWAYAM	Point-to-point messaging
6	SRM University	SRMSAT	Grating spectrometer
7	IIT, Kanpur	JUGNU	Near-IR camera
8	Consortium of Universities	STUDSAT	CMOS camera
9	Anna University	ANUSAT	Store - and - forward

3.2 DESIGN CONSIDERATIONS FOR PAYLOADS

Before discussing the details of these payloads, it is important to understand the following design considerations as applicable to each type of payload:

1. **Mass & size:** Typically, a Nanosatellite weighs about 10 Kgs and the payload weight can be only a fraction of this. This demands that the physical size and weight of the payload have to be compatible with the satellite bus.

2. **Power:** Electrical power for payload operation on a Nanosatellite is typically limited to less than 5W. This impacts payload designing in terms of frequency and duration of operation, volume of data generated and its storage/transmission.

3. **Data Collection/Transmission:** Usually, the International Telecommunication Union (ITU) allocates RF spectrum for Nanosatellites in VHF/ UHF (very high frequency/ultra high frequency) Amateur Radio band. (Some Nanosatellites use S-band for data downlink, with specific approvals). Due to this limitation, the volume of payload data gathering and transmission needs to be addressed.

4. **Attitude:** Accuracy of attitude pointing and stability depend on the payload requirements. For instance, an earth imaging camera pointing demands an accuracy better than +/- 0.1 degree and stability better than +/- 10e-3 degree/second.

5. **Antenna:** Some payloads may need specific antenna/probe to receive the desired signals. In case of volume constraint during launch, in-orbit deployment of such antenna/probe needs to be considered.

3.3 EARTH OBSERVATION PAYLOADS

Earth observation payloads mainly consist of optical or infrared cameras to image particular areas of interest. These cameras are generally operated from 'sun-synchronous' orbits to have a near-constant illumination of the earth's features. These cameras operate on the principles of 'remote sensing technology'. A few basics of these principles are explained in Annexure – A2. These basic principles drive the designing of a Nanosatellite.

3.3.1 Optical camera for earth observation

Evolving technology in the field of digital photography, mainly for mobile phones, resulted in miniaturised cameras that are most suitable for Nanosatellites. Most of these cameras operate with two dimensional CMOS (complementary metal-oxide semiconductor) detector array technology with ever-increasing packaging density in terms of 'megapixels' (MP). Driven by merits like parallel readout, pixel window selection, and binning of adjacent pixels, CMOS technology is preferred over the charge coupled device (CCD) technology, though the latter is superior in terms of noise performance.

A typical optical camera consists of an imaging lens assembly, a CMOS area array device in the focal plane, processing electronics and mechanical housing for mounting. The imaging lens will have either a fixed or variable F/number (ratio of focal length to Aperture diameter that controls the light flux entering the camera). The CMOS detector will have HxV number of pixels where H is the number of rows and V is the number of columns. These pixels are covered with a Bayer's pattern of RGB filters, which yield multicolour images using pre-defined algorithms (*see box*). This camera is operated in 'snapshot' mode with a pre-set exposure time which is limited by the maximum number of frames per second (FPS). It is important to adjust the exposure time, lens F/No and number of FPS so that the image is bright enough for data interpretation/ analysis. Also, these CMOS cameras will have a provision for an electronic shutter, operable in either 'global mode' or 'rolling mode'. The exposure time vis-à-vis the ground velocity should be adjusted to obtain an overlap between two consecutive images as shown in **Fig. 3.1.**

Each exposed frame generates a data volume of "number of pixels x B" bits where B is the quantisation number. For example, a camera with 1024 x 1024 pixels and 10-bit quantisation generates about 10 Mbits (megabits) data for each frame. A microprocessor with custom-built software controls the operation of the camera and an onboard memory (typical 2 GB capacity) stores the frame data for later transmission to the ground station. Two Nanosatellites developed by Indian academic institutions with such payloads are shown in **Fig. 3.2**

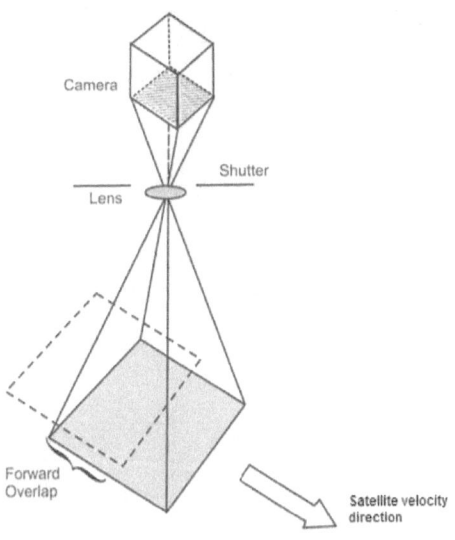

▲ **Fig. 3.1:** Overlap geometry for optical camera

NIUSAT
(Courtesy: spaceflight101)

PISAT
(Courtesy: Wikipedia)

▲ **Fig. 3.2:** Nanosatellites with an optical camera

Bayer's Filter

CMOS sensors in all digital cameras are covered with a Bayer's spectral Filter to produce natural colour images. Bayers's filter consists of a matrix arrangement of three primary colours – Red, Green and Blue in a regular pattern, each colour covering 'one pixel' of the CMOS detector. To get natural colour, combination of three colours is necessary – various algorithms are applied depending on the pattern of colours implemented. In this example, as per the Bayer's pattern shown in the Figure below, three colour contributions for a given pixel is as given below:

R11	G12	R13	G14	R15	G16	R17
G21	B22	G23	B24	G25	B26	G27
R31	G32	R33	G34	R35	G36	R37
G41	B42	G43	B44	G45	B46	G47
R51	G52	R53	G54	R55	G56	R57
G61	B62	G63	B64	G65	B66	G67
R71	G72	R73	G74	R75	G76	R77

For Pixel No.44,

Red $= (R33+R35+R53+R55)/4$

Green $= (G34+G43+G45+G54)/4$

Blue $= B44$

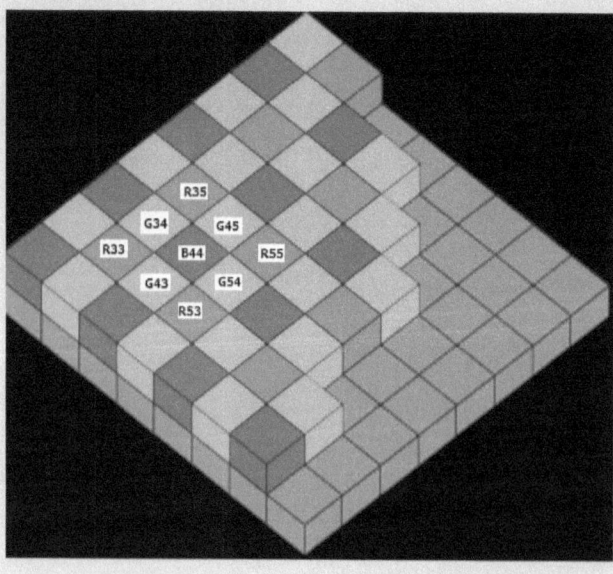

3.3.2 Infrared camera for earth observation

IR cameras are used for specific applications such as monitoring of vegetation moisture stress, snow-cloud differentiation, cloud top temperatures, forest/ bush fires and volcano studies. They operate in either shortwave infrared (SWIR) range of 1.5 – 1.75 μm, middle infrared (MIR) range of 3 – 5 μm or long-wave infrared (LWIR) of 7 – 14 μm range. Though the basic principles of IR cameras are similar to optical cameras, they normally provide coarser resolution due to low radiation intensity levels. While optical glass materials can be used for making optics for the SWIR range, special materials like calcium fluoride, zinc selenide and germanium are used for fabrication of MIR and LWIR Optics. Similarly, various combinations of Group III-V, IV-VI and II-VI materials are used for IR Detectors – such as indium gallium arsenide (InGaAS), lead telluride (PbTe), indium antimonide (InA) and mercury-cadmium-telluride (HgCdTe). Most of these IR detectors are operated at low temperatures (using thermo-electric coolers) to reduce thermal noise and improve the signal-to-noise ratio (SNR). IR detectors without coolers have also been developed recently. The important parameter to define the performance quality of IR cameras is 'noise equivalent temperature difference' (NEΔT)'. This is a measure of the minimum change in the radiant intensity level in the scene that can be detected by the camera and is expressed in milliKelvin at a given operating temperature. Nanosatellite 'Jugnu', developed by IIT-Kanpur, carried a near IR camera for earth imaging as shown in **Fig. 3.3**.

▲ **Fig. 3.3:** Jugnu Nanosatellite (Courtesy: IIT, Kanpur)

3.3.3 Spectrometers for earth observation

Spectrometers are more useful in the study of atmospheric pollutants, greenhouse gases, ocean colour and agricultural applications. While the optical and IR cameras operate with finite spectral bandwidths (typically 50 – 100 nm bandwidth), spectrometers operate with a continuous spectrum of a given spectral range. This continuous spectrum is generated by using either an optical prism, optical grating, linear variable filter or Fabry-Perot interferometer. The optical schematic of a spectrometer is shown in **Fig. 3.4.** The objective lens collimates the incoming light beam which is then dispersed into the spectrum. The imaging lens focuses this beam on the focal plane where a two-dimensional detector array is placed. While the spectrum is distributed in the velocity direction, the scene is scanned in the perpendicular direction such that each pixel is imaged in all wavelengths of the spectrum due to the forward motion of the satellite.

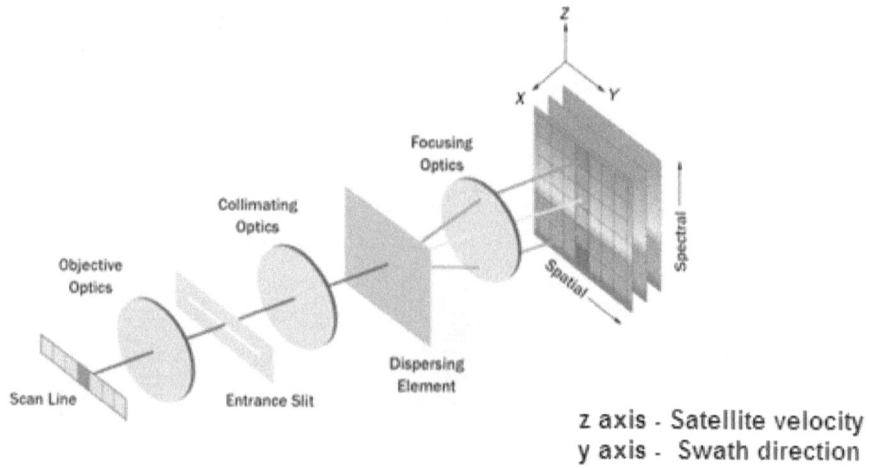

z axis - Satellite velocity
y axis - Swath direction

▲ **Fig. 3.4:** Optical schematic of a spectrometer (Courtesy: Semantic scholar.org)

3.3.4 Radio occultation sounders

These payloads are used for the study of earth's atmospheric temperature and humidity profiles, which are important for weather predictions. They operate on the principle that radio signals from GPS satellites passing through earth's horizon undergo refraction causing a bend in their path of travel which is a function of temperature and humidity. These radio signals from GPS satellites, which are either 'rising' or 'setting' near the horizon, travel through the atmosphere and are received by the GPS receiver onboard the Nanosatellite as shown in **Fig. 3.5**. The phase differences introduced in these received signals due to atmospheric bending are used to derive the local temperature and humidity and the relative motion between GPS satellites and Nanosatellites cover the vertical profile of atmospheric temperature. The position, velocity and time of the Nanosatellite at each instant of these measurements should be independently measured using a GPS/IRNSS receiver onboard the Nanosatellite.

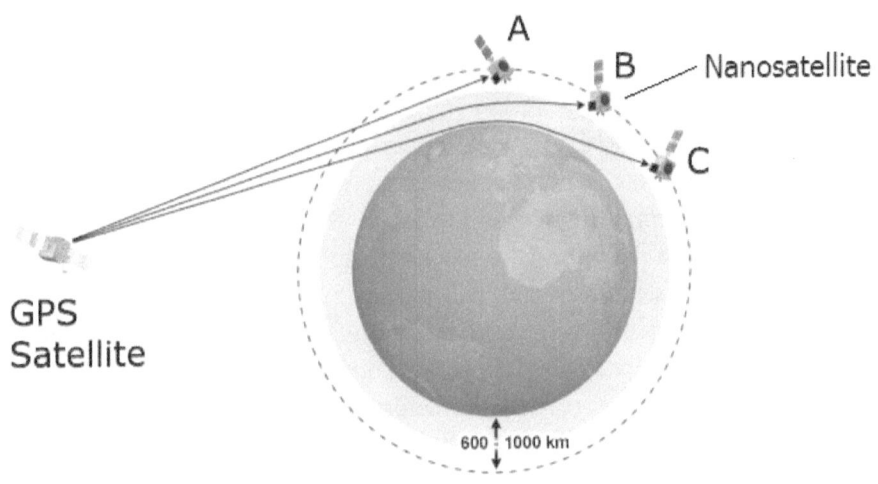

▲ **Fig. 3.5:** GPS Radio Occultation (Courtesy: UCAR)

3.4 COMMUNICATION PAYLOADS IN A NANOSATELLITE

Communication payloads in a Nanosatellite are mostly confined to RF signals at lower frequencies and low data rates due to the following reasons:

- Due to their size, weight and power limitations, high-frequency bands such as X-band and Ku-band are not employed in Nanosatellites.
- Frequency regulatory allocation to Nanosatellite is presently restricted to Amateur Radio bands in VHF/UHF range – though there are a few cases where S-band downlink is used for data transmission in these missions.
- Nanosatellites are mostly launched into LEO orbits (500 – 1000 kms) but not GEO orbits. Hence their usage for applications such as TV broadcasting is limited. (Hence they are not used for TV broadcasting and such applications.) However, recently a constellation of Nanosatellites (a

few 100s) has been launched into LEO orbits for communications such as internet services by a US company, OneWeb.

3.4.1 Store and forward systems

The S&F (store and forward systems) payload is a simple communication payload for a Nanosatellite and it uses VHF and UHF radio links for receiving and transmitting digital messages such as text, voice or email. A message from one ground terminal is uplinked and stored in the memory onboard the satellite and transmitted as the satellite passes over the designated ground terminal. It can also be used to collect message data from unattended platforms in inaccessible places such as weather stations and transmit them to the required destination. The two ground terminals do not have to see the satellite at the same time and a single satellite can provide global communications coverage. Data transmission on the downlink (from satellite to user terminal) is performed in the time-division multiplexed (TDM) mode. Multiple access to receiving channel of the onboard transceiver (uplink) is performed according to split channel reservation multiple access (SRMA) protocol. Phase shift keying (PSK) modulation is used for downlink and uplink transmissions. A typical amateur radio protocol such as AX.25 is used for communications. In order to enhance the performance of satellite communications, link and error correction techniques are used in this payload.

3.4.2 Study of ionospheric effects on RF communications

The ionosphere is a region of the upper atmosphere that extends from about 70 kms to 500 kms in altitude. It is a region where the atmosphere is partially ionised due to extreme ultraviolet and X-ray radiations from the sun - hence the name, ionosphere. This ionised gas, also referred to as 'plasma', will interact with electromagnetic signals (such as radio waves) that pass through it. The plasma layer will reflect radio signals below 30 MHz but will allow the propagation of high frequency (HF) signals around the globe. The density of ionospheric layer changes with the time of day, altitude, latitude, season and solar activity. The major effects of the ionosphere on radio waves include:
- Dispersion, where signals of different frequencies travel at slightly different velocities which causes signal delay and differential delay in wideband communication systems.
- Refraction (bending) is an important consideration for radars which track space objects as it causes them to see the object in a position displaced from the true position.

- Faraday rotation effect - when a linearly polarised signal propagates through a plasma in the presence of a magnetic field, the plane of polarisation rotates. The amount of rotation is proportional to the magnitude of the magnetic field and the total ionisation (total electron content, TEC) through which the signal passes. It is also inversely proportional to the square of the frequency of the signal. At VHF frequencies and high solar activity levels, the plane of polarisation may be rotated through many times of 360 degrees. At C-band frequencies (4 GHz) the signal rotation will rarely exceed a few degrees.

Ionospheric effects on the navigational satellite signals (such as GPS, IRNSS, GLOSNAS) cause delays in phase and pseudo ranges, which in turn result in inaccuracy in position determination.

Total electron content (or TEC) is an important descriptive quantity for the ionosphere of the earth, which causes above mentioned effects on radio wave propagation. TEC is defined as the total number of electrons integrated between two points, along a tube of one meter squared cross-section, i.e., the electron columnar number density. It is expressed in multiples of the so-called TEC unit, defined as 1 TECU=10^{16} el/m^2. Measurement of TEC at a given location at different times of a day is very useful to characterise the behaviour of RF signals and minimise transmission induced errors. If this measurement is done from different geographic locations, a 'TEC Tomographic map' can be generated.

TEC from a Nanosatellite can be measured using the following techniques:
- Transmit two plane polarised radio waves in UHF band from the Nanosatellite with accurately known frequencies and polarisation and measure the Faraday rotations in the ground received signals using a Yagi antenna
- Transmit three equally spaced radio signals from the Nanosatellite (f_0 – f_m, f_0, f_0+f_m) and measure the differential delays for the three signals. At the ground station, the signals with three frequencies are separated, the terms with centre frequencies of f_0 - fm and f_0 + fm are mixed with the f_0 signal, respectively. The resulting two signals after mixing both have a centre frequency of fm. The phase difference between these two signals is a measure of TEC.

'Pratham', the Nanosatellite developed by IIT, Bombay which carried a payload to measure TEC in the ionosphere, is shown in **Fig. 3.6.**

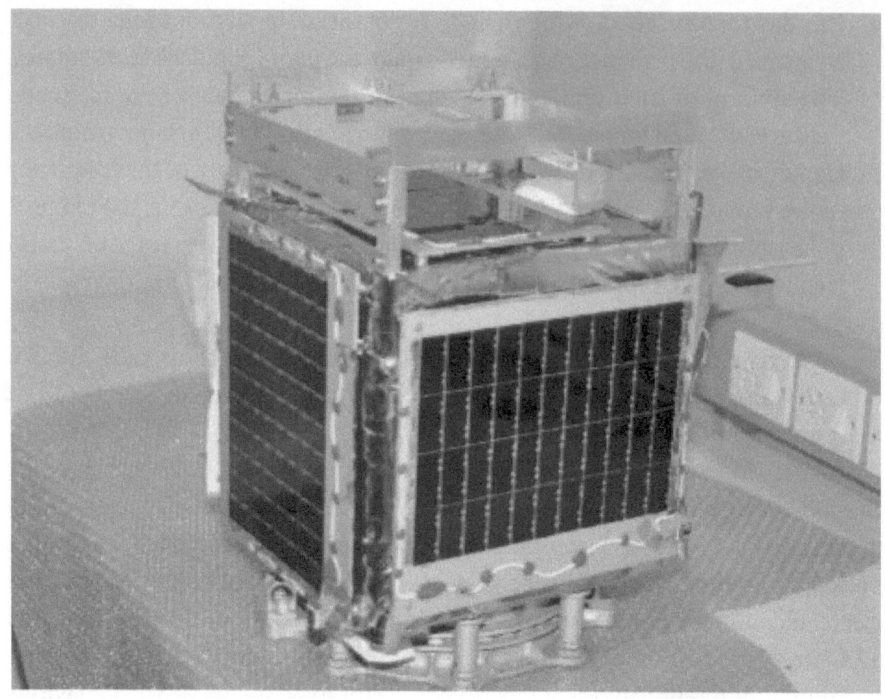

▲ **Fig. 3.6:** Pratham Nanosatellite (Courtesy: IIT, Mumbai)

3.4.3 Automatic identification of ships (AIS)

The AIS payload is used for identification of ships. This payload consists of a two-band radio receiver onboard a Nanosatellite to receive the message signals transmitted by ships and decoded to identify their details. As per International Maritime Organisation rules, all ships (cargo and passenger) have to transmit periodical messages giving details about their unique ID number, country of origin, destination, heading, speed and direction of travel as well as location by GPS to coastal stations. These signals from ships near the horizon are picked up by a Nanosatellite with AIS payload to identify ships using complex algorithms.

The AIS messages are transmitted by ships in two VHF frequencies, 161.975 MHz and 162.025 MHz at 9.6 kbit/s, employing Gaussian minimum shift keying (GMSK) modulation over 25 kHz channels using the high-level data link control (HDLC) packet protocol. From the satellite orbit, a large number of ships will be in its 'radio footprint' and there will be an overlap or 'collision' of AIS messages from different ships.

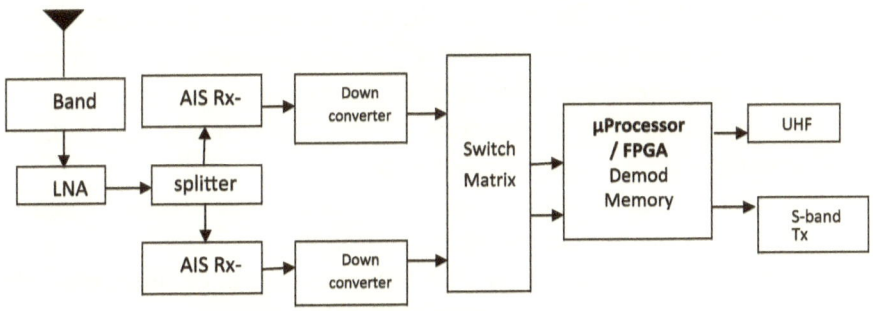

▲ **Fig. 3.7:** Typical block schematic of AIS receiver

However, as these signals are transmitted from each ship in a 'self-organised time division multiple access' (SOTDMA) scheme where information on the TDMA slot for the next message is transmitted *'a priori'*, each TDMA slot is time synchronised with the GPS time. Resolving the time correlation of each received signal and decoding the received messages is the real challenge in this system.

A typical block schematic of AIS payload is shown in **Fig. 3.7.** It consists of a monopole VHF antenna, low noise amplifiers, two VHF radio receivers and a programmable FPGA based signal processor. The pre-processed data is stored in the onboard memory and transmitted to the ground station in UHF or S-band.

3.4.4 Laser communication payload

As mentioned earlier, the major limitations of a Nanosatellite RF communication system are low-frequency operations with lower data rates and regulatory restrictions in the allocation of higher frequency bands. One way of avoiding both these limitations is to use an optical communication link, which makes a large bandwidth available for data transmission without international regulation on their usage. An optical communication link may employ either a laser beam or an array of light-emitting diodes (LED) with a relative trade-off between them.

The basic elements required for an optical communication link are a transmitter of light beam carrier with collimating optics, electro-optical modulator to modulate the signal on to the carrier, a receiver consisting of an optical telescope with a quadrant avalanche photodiode (APD) detector, a demodulator and signal processing electronics. In case of two-way communication, that is, satellite to ground and vice-versa these elements are required to be installed both on the satellite and on the ground station. However, it may be noted that one-way link from a Nanosatellite to a ground station is more preferable due to the following reasons:

- Satellite receiver telescope aperture has to be small in size due to the size and weight limitations of a Nanosatellite
- Attitude pointing of Nanosatellite will not be accurate enough to point the onboard receiver towards the ground transmitter
- Volume of data transfer from satellite to ground will be much more than ground to satellite (mostly telecommand data only)

The source of light can be a laser beam of suitable wavelength to meet the criteria such as atmospheric transmittance, availability of detectors and other components and also eye safety guidelines. As visible light wavelengths are prone to atmospheric scattering and background noise, lasers operating in the near-infrared (NIR) wavelengths (800 – 1550 nm) are preferable. While the laser beam with a smaller angle of divergence gives better SNR and modulation bandwidth demands on a Nanosatellite, the ADCS will be more stringent to point the beam to the ground receiver. LEDs offer an alternative choice with large beam width thereby reducing the SNR and modulation data rate. To modulate the digital data on to the optical carrier, two types of intensity modulation (IM) techniques are usually employed, namely, on-off-keying (OOK) and pulse position modulation (PPM).

The OOK modulation uses the intensity variations of the transmitter to encode bits and can be recovered with clock recovery units. But in the OOK modulation, it is important for the receiver to identify a suitable threshold between "ON" and "OFF" states particularly when the signal is corrupted with noise and stray light interference. In the PPM scheme, data is encoded into symbols consisting of N slots and each symbol is encoded to $\log_2 N$ bits per symbol. The main advantage of PPM over OOK is that it is "self thresholding" (the receiver simply chooses the highest intensity slot) and that it has a low duty cycle, which can be advantageous in average-power-limited transmitters.

3.5 SCIENTIFIC PAYLOADS

A Nanosatellite platform can be effectively used by academic institutions to conduct some short-term scientific experiments or observations to demonstrate a scientific hypothesis or a concept. These payloads may be evolved in collaboration between scientific research laboratories and institutes of engineering and technology.

3.5.1 Biological experimental payloads

Towards future human settlements on other planets, there is always interest to know how various biological processes behave in space environment under near-zero gravity. It may be survival and behavioural growth of biological cells, germination and growth of plant species or evolution of genes. Several payloads have been flown and under development by NASA and other agencies.

- **PoreSat** is an autonomous, free-flying spacecraft that will investigate how germinating plant cells sense and respond to gravity. Researchers are studying spores in space to gain a more detailed understanding of molecular and biophysical mechanisms for gravity sensing. Specifically, it will investigate how germinating single-celled spores of the aquatic fern *Ceratopteris richardii* sense and respond to gravity

- **Micro-7** is the first satellite to study gene and microRNA expression in non-dividing cells. The study will also investigate how spaceflight affects the response of non-dividing cells to DNA damage. The data from Micro-7 will provide insight into how gene expression regulates cellular adaptation to spaceflight and the specific role of microRNA in this process

- **GeneSat-1** is a fully-automated, miniaturised satellite that provides life support, nutrient delivery and performs assays for genetic changes in E. Coli - that's the abbreviated name of the bacterium Escherichia (Genus) coli (Species).

A miniaturised camera will continuously monitor these experiments and send the data to the ground station. Many of these experiments require stringent control of ambient conditions such as temperature, pressure and humidity.

3.5.2 Monitoring of earthquakes

Due to the large impact on humans and other living beings as well as property, research has been going on to predict earthquakes. Several hypotheses were evolved to decide the 'precursors' for the occurrence of an earthquake event and experiments carried out to establish these hypotheses. With the launch of European Sentinel-1, which carried a synthetic aperture radar (SAR), precise movements in the earth's crusts could be measured using SAR in the interferometric mode. There are two methods employed in satellite-based earthquake prediction studies as given below:

- In the first method, extremely low frequency (ELF) magnetic signal data is detected and recorded which may lead to the prediction of earthquake activity. According to the theory, fracturing bedrock along the fault lines create ELF magnetic waves. These waves radiate from the earthquake's hypocentre region (several tens of km²), through the earth (5-80 kms), through the atmosphere to the ionosphere (100-200 kms), and are propagated up the earth's magnetic field lines to altitude. The 'Quakesat' developed by Stanford University, the USA is a Nanosatellite, carrying a payload comprising a single axis magnetometer and a small E-field dipole antenna. The objective of this satellite is the detection of ELF signatures and the strategy used to detect the small magnetic field measurements (0.5 to 1000 Hz) relies on the use of a very sensitive AC magnetometer. The ELF sensor is a single axis, search coil (induction) type magnetometer with multiple frequency bands (0.5 – 10 Hz, 10 – 150 Hz, 10 – 1000 Hz and 127 – 153 Hz band).

- The second method relies on the hypothesis that during the formation of an earthquake, ultra low frequency (ULF) (f <5 Hz)/extra low frequency (ELF) (f = 30 -300 Hz) electromagnetic waves are observed on the ground and low earth orbit altitudes. These seismo-electromagnetic emissions travel through the atmosphere and are captured near the ionosphere-magnetosphere transition region and continue their journey along the geomagnetic field lines as Alfven waves. These waves, near the inner Van Allen radiation belt boundary, interact resonantly with the trapped particles (protons, electrons) in the radiation belt and cause their precipitation. This precipitation may be observed a few hours before the manifestation of an earthquake by a satellite in the LEO as particle bursts (sudden increase in the particle counting rates). IIT, Madras has developed a Nanosatellite carrying a payload to detect charged particles using scintillation-based techniques.

CONCLUSION

With the recent technology developments in miniaturisation of electronics and sensor devices, there are several opportunities for academic institutions to develop various payloads to be flown on Nanosatellites. These payloads can be used, i) for a specific mission objective such as the study of a coastal/forest area in the region; ii) to validate a scientific hypothesis; or iii) to demonstrate a new technology which can be adopted for operational missions. This exercise will enable the student community to gain hands-on experience in satellite technology and project management activities.

REFERENCES

1. *'Earthquake Monitoring Gets Boost from New Satellite'* by John R. Elliott, Austin J. Elliott, Andrew Hooper, Yngvar Larsen, Petar Marinkovic, and Tim J. Wright, Earth & Space Science News
2. eoPortal Directory – Quakesat
3. *'A Nanosatellite Mission to Study Charged Particle Precipitation from the Van Allen Radiation Belts caused due to Seismo-Electromagnetic Emissions'* by Nithin Sivadas, Akshay Gulati, Deepti Kannapan, Ananth Saran Yalamarthy, Ankit Dhiman, Arjun Bhagoji, Athreya Shankar, Nitin Prasad, Harishankar Ramachandran and R. David Koilpillai, Indian Institute of Technology Madras
4. *'Genesat-1: Small satellite tackles Big Biology questions'* by Leonard David, Space.com's Space Insider Columnist, August 30, 2005
5. NASA Ames launching Nanosatellites, Science Experiments on SpaceX Rocket, April 10, 2014
6. eoPortal Directory – Galassia
7. *'Pratham: Faraday Rotation-Based TEC Measurement'* by Giri Prashanth, Avnish Kumar and Sanyam Mulay, Indian Institute of Technology Bombay, Indian Journal of Radio & Space Physics Volume 42, June 2013, pp 197-203
8. *'Store and Forward Communication Payload Design for LEO Systems'* by Dariush Abbasi-Moghadam, Mojtaba Abolghasemi, Majlesi Journal of Electrical Engineering, Vol. 10, No. 3, September 2016

* * *

M. Venkata Rao has more than 30 years of experience in system engineering aspects of electro-optical systems and earth observation payloads and project management of Indian remote sensing satellites program. He has served as Project Director for two of ISRO's Remote sensing satellites.

NANOSATELLITE STRUCTURE

R Arunachalam

A structure is the foundation and building edifice on which the satellite bus and payload systems are integrated to form a monolith of any satellite. Conventional spacecraft sub-systems are designed and manufactured separately and are integrated on to the structure during various stages of satellite development.

Structural materials used for space application have evolved over decades for meeting the requirement of stringent design, covering factors of strength, lightweight and optimal volume to make the satellite practicable and cost-effective for launch. Also, the structure has to withstand the stresses at different phases of development, like the quasi-static and dynamic loads/stresses faced by the satellite during fabrication phase, satellite transportation and launch phase, till injection into the orbit.

Structural design, analysis and fabrication are unique combinations of mechanical and aeronautical engineering embedded with material science and process engineering. While these disciplines are continuously evolving with the development of aerospace technology, the present-day Nano and Microsatellites are true beneficiaries in the form well-optimised CubeSat structure, which is the order of the day

The reader will get a good insight into these aspects in this chapter.

The main functions of a satellite structure are: (i) to accommodate sub-systems; (ii) to withstand launch loads and space environment; (iii) to provide required orientation for various sub-systems; (iv) to provide ease of accessibility (for assembly and integration); and (v) to provide good electrical ground and electromagnetic compatibility.

Aerospace structures generally require lightweight designs with the main goal of achieving higher strength to weight ratio. Satellite structural designs have been evolved over the years with the development of lightweight materials such as aluminium honeycomb core and reinforced plastic. A brief description of the design aspects of satellite structures in general and Nanosatellite structure, in particular, is covered in the following sections.

4.1 TYPES OF STRUCTURE DESIGNS

Primary structures are designed basically using the following methods:
- Skin-frame structures
- Truss structures
- Monocoque cylinders
- Skin-stringer structures

4.1.1 Skin-frame structure

The skin-frame structural design uses an interior skeletal network of axial and lateral frames to mount exterior skin panels using fasteners or rivets. The frames support bending, torsion and axial forces. The skin reinforces the structure by supporting the shear forces introduced by the interior member connections. The skin's thickness is sometimes minimised to save mass but excessively thin skin leads to structural instability. A typical skin frame structure is shown in **Fig. 4.1.**

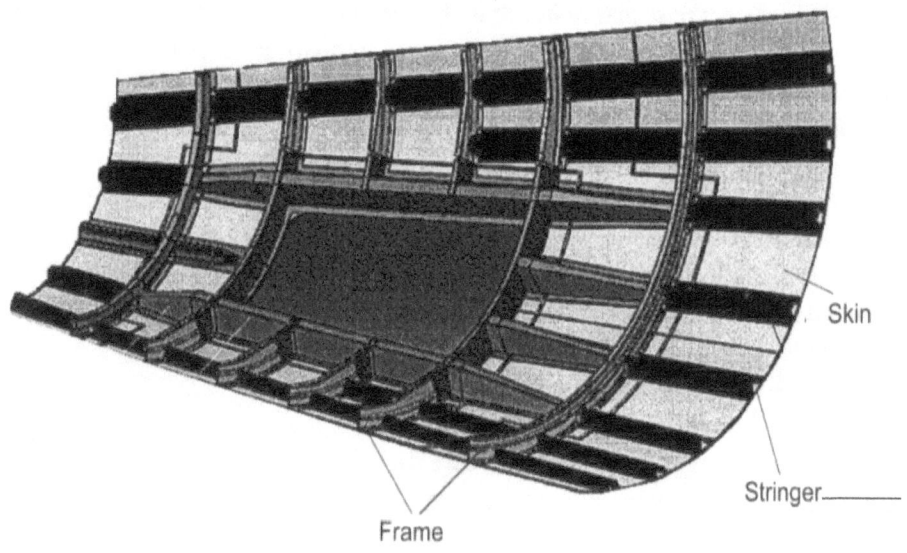

▲ **Fig. 4.1:** Skin frame structure (Courtesy: aerospaceengineering.net)

4.1.2 Truss structures

Truss structures use an array of members that can only support axial loads. Truss members are produced independently and arranged typically in arrays of triangles for stability. The members are manufactured using extruded tubes made of composite, metallic or sheet metal materials. A stable truss is statically determinate and has no excess members to introduce alternative load paths. Trusses are generally mass efficient components, mounted both internally and externally and the absence of shear panels enables easy access to a payload. However, the absence of shear panels is not helpful to a spacecraft with body-mounted solar panels. **Fig. 4.2** shows a truss type structure.

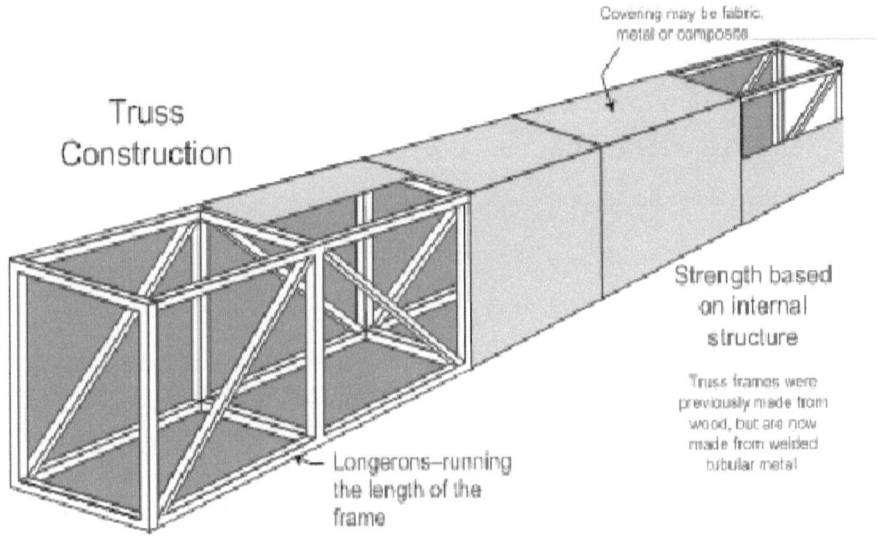

▲ **Fig. 4.2:** Typical truss based structure (Courtesy: reprap.org)

4.1.3 Monocoque Cylinder

Monocoque cylinders are axisymmetric shells that do not contain stiffeners or frames. The shells are manufactured using metallic or sandwich panels with curved sections formed by rolling. Typically, two or three curved sections are fabricated and assembled into the cylindrical configuration. A typical Monocoque structure is shown in **Fig. 4.3.**

The strength of the monocoque cylinder is usually limited by its buckling strength. The shells are most efficient when the loads are distributed evenly throughout the structure. Components are typically mounted to the walls using fasteners; however, care must be taken not to overload the shell and cause local failures. The monocoque cylinder design is compatible with a spacecraft with body-mounted solar cells and relatively lightweight components.

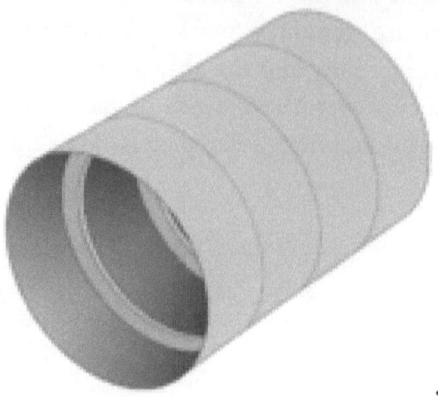

▲ **Fig. 4.3:** Monocoque Cylinder (Courtesy: https://www.wikiwand.com/en/Airframe)

4.1.4 Skin-stringer structure

Cylindrical skin-stringer based structures are designed using axial and lateral frame members attached to an outer skin. The skin's thickness is optimised to avoid structural instability. Typical connection methods include fasteners and/or rivets. Interior components are usually mounted to the walls at locations along the stringer assembly. The skin must be designed sufficiently stiff to enable proper mounting of exterior entities such as body-mounted solar cells. A typical skin-stringer based structure is shown in **Fig. 4.4.**

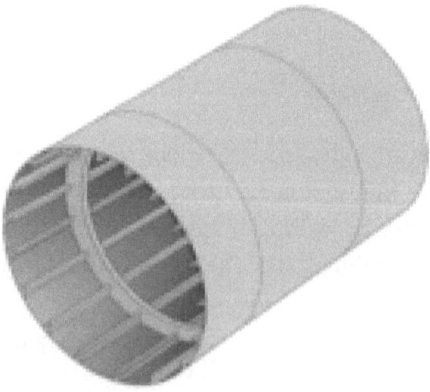

▲ **Fig. 4.4:** Skin-stringer structure (Courtesy: https://www.wikiwand.com/en/Airframe)

4.2 BUILDING OF STRUCTURE

New techniques such as multifunctional structure, improved composite construction and monolithic designs are some of the options used for building Nanosatellite structures to improve efficiency beyond the current level apart from adhering to basic philosophy.

Once the configuration of a Nanosatellite is defined, the primary structure is sized to comply with the strength, stiffness and safety requirements. Several methods are available to optimise the structural properties of a satellite. The optimum method may vary depending on the design. The three most widely used methods are:
- Sandwich structures
- Multifunctional structures
- Isogrid structure

4.2.1 Sandwich structure

Sandwich structures are often used in skin-frame designs. A sandwich structure consists of two thin face sheets attached to both sides of a lightweight core. The design of sandwich structure allows the outer face sheets to carry the axial loads, bending moments and in-plane shears while the core carries the normal flexural shears.

Sandwich panel face sheets are commonly fabricated using aluminium or graphite/epoxy composite panels. The core is typically fabricated using a honeycomb or aluminium foam construction. Honeycomb sandwich panelling is the lightest option for compressive or bending load applications. The disadvantages of using honeycomb cores are that they require potted inserts for mounting, which results in thermal inefficiencies. The low thermal conductivity of the adhesive layers used in honeycomb construction makes them prohibitive in optical mirror mounting aerospace applications. A typical construction of the sandwich structure is shown in **Fig. 4.5** and an actual sample is shown in **Fig. 4.6**.

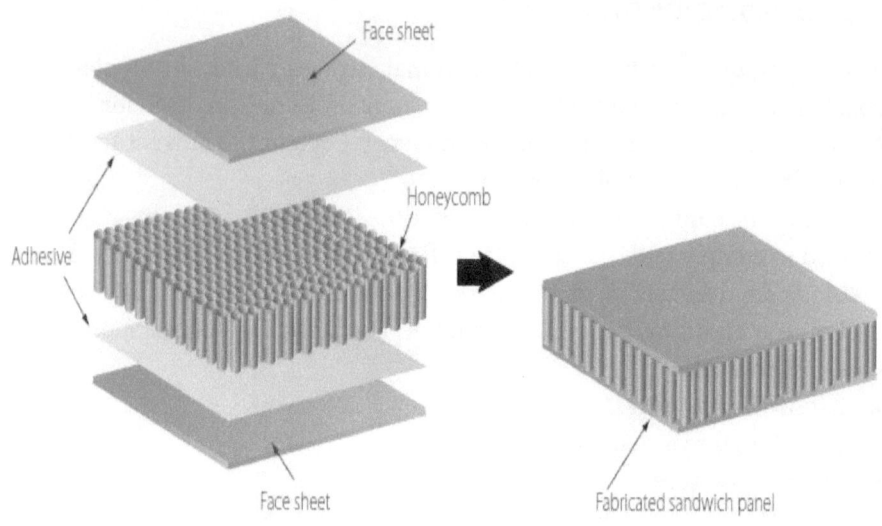

▲ **Fig. 4.5:** Sandwich Construction
(Courtesy: Tasuns Composite Technology Co. Ltd)

▲ **Fig. 4.6:** Sandwich construction with CRP skin
(Courtesy: PROTECH Composites, Vancouver, WA, USA)

4.2.2 Multifunctional structures

Conventional spacecraft sub-systems are designed and manufactured separately and are integrated only during the final stages of the satellite's development. This requires a separate package for each sub-system hardware with associated mechanical interfaces, panels, frames and bulky wire harnesses, which add considerable mass and volume. As all sub-systems are generally secured to the structure - the multifunctional structure (MFS) design aims at merging these elements into the main structure - so that the main structure also carries out some of the typical functions of the sub-systems (e.g. electrical energy storage).

The main advantages are: (i) removal of the bolted mechanical interfaces and most of the sub-system packages; (ii) reduction of satellite structure mass as the strength of the parts of the sub-system embedded into the structure are exploited and substitute purely structural parts; (iii) reduction of the overall satellite volume, as elements such as battery packs or electronic harnesses can be built into the structure volume. Miniaturised multi-chip module (MCM) electronic devices and flexible circuit boards can be directly mounted on this structure. A schematic MFS is shown in **Fig. 4.7**.

The design allows for an easily accessible, removable, and modular electrical system. The benefits of this technology include a 70 per cent reduction in electronic enclosures and harnesses and a 50 per cent reduction in spacecraft volume required for mounting these conventional packages.

A sandwich structural panel consists of an aluminium honeycomb core and lightweight CFRP face sheets. Integration of electronics is implemented within the panel by mounting electronics on a multi-layered composite enclosure with multi-materials. This composite enclosure provides load-bearing, effective thermal conduction, radiation shielding capabilities and space for embedding electronics.

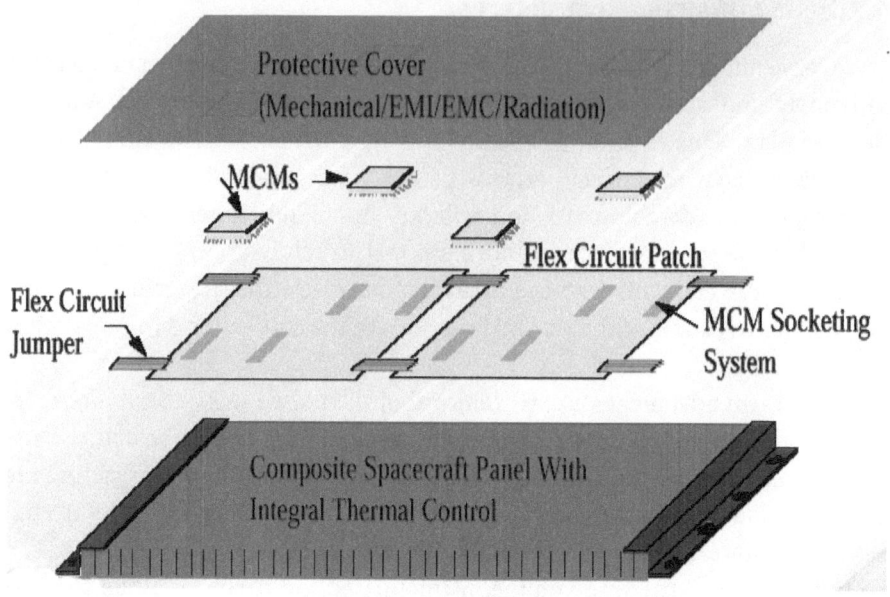

▲ **Fig. 4.7:** Schematic of a typical Multifunction Structure
(Courtesy: Lockheed Martin eoPortal Directory)

4.2.3 Isogrid structure

An isogrid is a type of partially hollowed-out structure formed usually from a single metal plate (or face sheet) with triangular integral stiffening ribs (often called stringers). It is extremely light and stiff. Compared to other materials, it is expensive to manufacture and so it is restricted to spaceflight applications and some particularly critical parts in general aerospace use. It is isotropic in nature for in-plane loading. Isogrid structures are lightweight and offer high specific strength, bending and buckling strength.

An isogrid uses an array of equilateral triangle cut-outs to increase the stiffness per weight of a structure. The pattern may be manufactured by machining a metallic panel or it may be constructed using composite materials. The concept began with using metal structures and the development continues focusing primarily on composite applications. Two types of isogrid structures are shown in **Fig. 4.8.**

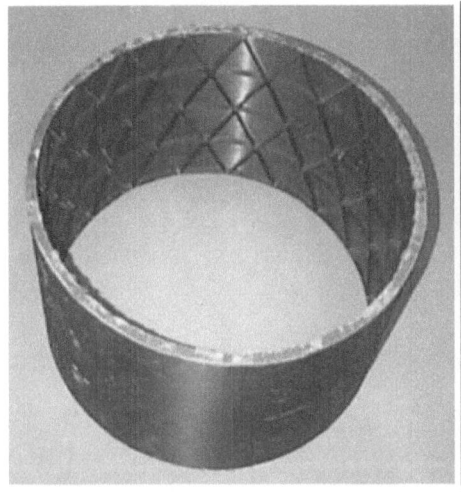 

▲ **Fig. 4.8(a):** Metallic isogrid structure (Courtesy: http://citeseerx.ist.psu.edu/)

Fig. 4.8(b): CFRP isogrid structure (Courtesy: https://www.iccm-central.org/)

4.3 CUBESAT STRUCTURE

The CubeSat design was first proposed in 1999 by Professors Jordi Puig-Suari of California Polytechnic State University and Bob Twiggs of Morehead State University. Their goal was to enable graduate students to be able to design, build, test and operate a small satellite in space. Although initially, only universities used this design, CubeSat became a standard for Nanosatellites and more than 500 such satellites have been launched. The basic unit (1U) of a CubeSat is 10x10x10 cm³ in size weighing 1.33 kg and multiples of this units can be stacked like 2U, 3U up to 16U as shown in **Fig. 4.9**.

CubeSat sub-systems, including its primary structure, are standardised and commercialised such that anyone can buy and assemble the same. Pumpkin Inc, the USA supplies CubeSat structure based on Monocoque design and fabricated with aluminium alloy Al5052-H32 with a solid wall or skeletonised wall. There are other suppliers of CubeSat structure such as Clyde Space and Endurosat.

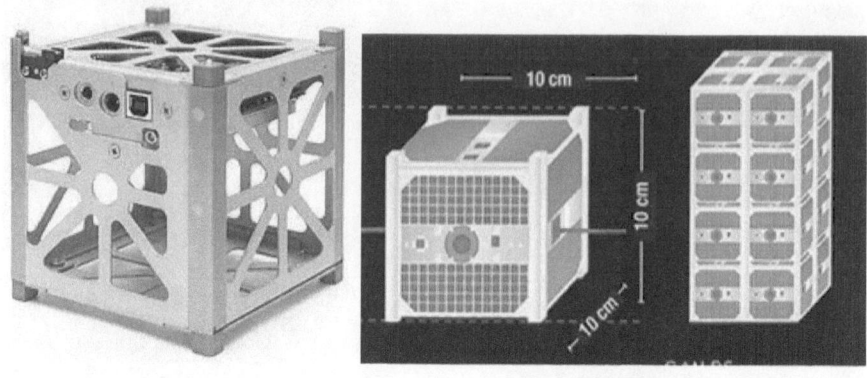

(a) One Unit (1U) (b) Maximum16 Units
(Courtesy: https://en.wikipedia.org/wiki/CubeSat) (Courtesy: Canadian Space Agency)

▲ **Fig. 4.9:** CubeSat structure

4.4 MATERIALS FOR STRUCTURE

For manufacture of lightweight structures, mostly alloys of aluminium, titanium, and stainless steel, as well as beryllium and composite materials are used.

4.4.1 Aluminium alloys

Aluminium alloys are the most widely used metal in spacecraft manufacturing for its high strength-to-weight ratio, high ductility and ease of machining. Although the stiffness-to-weight ratio is comparable to steel, the strength to weight ratio is typically higher. The disadvantages include low hardness and a high coefficient of thermal expansion. These alloys are tempered to increase their material strength. Two typical alloys used in manufacturing these are - Al6061-T6, containing silicon and magnesium which strengthen the alloy during tempering. This alloy has good machinability and corrosion resistance; and Al7075-T7 alloy, containing zinc and traces of magnesium. The alloy exhibits higher strength than Al6061-T6 but is more difficult to machine.

4.4.2 Titanium alloys

Titanium and titanium alloys are used for applications requiring very high strength. These materials exhibit high strength-to-weight ratios, low coefficient of thermal expansion and excellent corrosion resistance. Ti-6Al-4V, which contains six per cent aluminium and four per cent vanadium is the most popular titanium alloy used in aerospace applications.

4.4.3 Stainless steel alloys

Steel is mainly used in aerospace applications where low-volume strength and stiffness are important. Steel provides high wear resistance. Austenitic stainless steel is mainly used in spacecraft. It contains 12 per cent chromium, which results in a tough chromium-oxide coating that protects parts from corrosion. Stainless steel is non-magnetic and certain low-carbon alloys can be welded without sensitisation. The metal is generally used for fasteners.

4.4.4 Beryllium

Beryllium is used for very high stiffness aerospace applications. It has a specific modulus which is 6.2 times that of aluminium. The material is non-isotropic due to its grain alignment and therefore exhibits low ductility and fracture toughness in its short-grain direction. It is commonly used in lightweight optics and mirrors because it performs well at cryogenic temperatures (i.e. low CTE and high thermal conductivity). However, beryllium is expensive, difficult to machine and sparsely available. Beryllium must be machined in a controlled environment because its powder, inhaled is hazardous. The parts need to be safely handled once machined.

4.4.5 Composite materials

Composites are unique, combining two or more materials, utilising the respective advantages of the participating materials. Composite structures consist of a matrix and reinforcement. The matrix (metal, epoxy) binds the reinforcing materials (carbon, graphite) together into a continuous system. The efficiency of composite structures is due to its high specific modulus. The flexural shear loads are transferred from the matrix to axial loads on the high-strength fibres, creating a structure three to five times stiffer than aluminium at 60 per cent of the mass. Discontinuous-reinforced composites comprise ceramic or fibre particles that are randomly distributed throughout the matrix. Aluminium, reinforced with silicon carbide particles, is the most widely used discontinuous composite.

Majority of continuous fibre composites are generally called laminate composites. These laminate composites are manufactured from several layers of woven fibres called laminae. The laminae are composed of several parallel fibres arranged in sheets as shown in **Fig. 4.10**. Stacking several of the laminae with fibres aligned at different angles, called lamina angles, creates a more stable laminate composite structure. The laminate may be customised for individual applications by varying the fibre type and the layup. For example, some graphite/epoxy laminates are tailored to have a nearly zero CTE and others may be laid up to exhibit extraordinary specific stiffness properties.

Although composite materials offer many advantages over the metals, especially low density and high strength, they involve complex manufacturing procedures and processes, which lead to lengthy and costly development time. Also, the anisotropic nature of the material requires special care when load paths are calculated and special methods of attachment have to be utilised (adhesives or fasteners).

However, this disadvantage can be absorbed if a modular construction approach is implemented and advanced manufacturing methods are applied. The overall manufacturing cost increase is only a fraction of the total development cost and this increase is offset by the overall reduction in the structural mass.

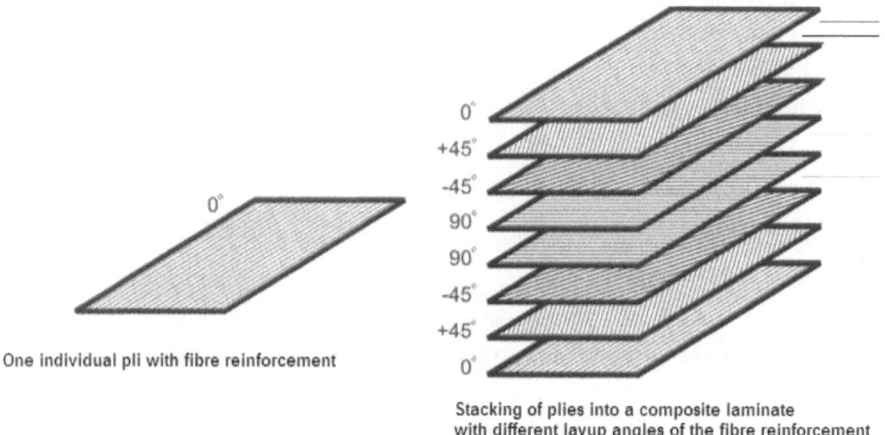

One individual pli with fibre reinforcement

Stacking of plies into a composite laminate
with different layup angles of the fibre reinforcement

▲ **Fig. 4.10:** Composite laminate fabrication
(Courtesy: Dr. Bicerano's blog on polymer and composite materials science)

4.5 STRUCTURE ANALYSIS

The analysis of a satellite's structural design is done using finite element method to verify whether the design of all structural elements can withstand the launch loads, whether the stresses and deformations are within the strength of the materials used and to ensure that enough safety margins exist.

Finite element analysis is achieved by dividing the structure under examination into small units or "elements". The elements can be given various shapes, but rectangles and triangles are most commonly used. These elements are connected through the "node" points.

The accuracy of calculations, therefore, depends greatly on the quality of the modelling of the structure, assumptions made for boundary conditions, the knowledge regarding loading conditions and environment and the number as well as type of elements used to represent the structure. Although finer quality of elements can accrue greater accuracy of calculations, it involves longer computation time, resulting in increased costs. So, the accuracy of the analysis should be judiciously decided to balance the cost and time factors.

4.6 TESTING OF STRUCTURE

Normally a satellite's structure undergoes a static test for qualifying the design and manufacturing process of structure for quasi-static loads experienced by the launch vehicle at the time of launch using a separate structural model. Since the structure of a Nanosatellite is small, a separate static test is not conducted on the structure. However static load and dynamic load vibration tests are conducted on the fully assembled satellite using an electro-dynamic shaker as shown in **Fig. 4.11**.

Simulation of a quasi-static load test is carried out for the critical load case which is selected by combining all load cases for the launch vehicle. Sinusoidal and random vibration tests are conducted as per the load specification given by the launch vehicle data agency.

The integrity of the structure/satellite is verified by comparing the pre and post vibration test responses, visual inspection of critical locations/ joints and also by conducting electrical tests of the satellite after the vibration test.

▲ **Fig. 4.11:** A satellite mounted on the shaker for vibration test (Courtesy: ISRO)

✳ ✳ ✳

R. Arunachalam is a quality and reliability expert for the mechanical systems for various satellites of ISRO. His specialisation is quality assurance of satellite structures and mechanical elements, including deployment mechanisms and test evaluation of Satellite mechanical systems.

THERMAL CONTROL SYSTEM

R. Arunachalam

Any satellite in orbit is always immersed in the space environment consisting of deep cold space, solar radiation, the sun's radiation as reflected by the earth (termed as earth albedo), earth's emitted heat (termed as infrared radiation), all impinging on the satellite in unison. This results in uncontrolled heating/cooling of the different faces of a satellite cyclically as per the orbital revolution and the season. The heat generated internally by the bus and payload systems further adds up to the external radiation

Since the cold space environment is an all-pervasive vacuum without atmospheric air around, the convection mode of heat transfer is totally absent. The only mode of heat transfer in space is primarily by radiation along with additional conduction mode through the satellite sub-system's body contact surfaces.

Thermal engineering is a specialised branch of mechanical engineering/ thermodynamics covering design, simulation, analysis and test. In this effort the heat within the satellite is judiciously controlled irrespective of diurnal or seasonal variations due to external factors, thereby maintaining acceptable and safe temperature limits for the satellite systems.

The reader will get a good understanding of this engineering science in this chapter.

The main function of a satellite's thermal control system (TCS) is to keep the temperature of all sub-systems within their specified ranges during all mission phases. It needs to maintain a thermal balance between inputs from the external environment, which can vary widely as the satellite is exposed to cold deep space, solar radiation and earth's albedo (reflected sunlight) and shine (emitted IR radiation) and the internal heat generated during the satellite's ejection.

Thermal control is essential for the success of a mission because the temperature varying too high or too low could damage the performance of components and consequently affect the mission. Thermal control is also necessary to keep temperatures of components such as batteries, optical sensors and atomic clocks within specified limits. Typical temperature specifications of sub-systems are shown in the table below.

The low mass of a Nanosatellite means low thermal capacitance and faster thermal response. The small size can also result in high power densities (power per unit volume), particularly in a high mission utility operational satellite.

Sl No.	Sub-system	Typical Temperature range
1	Li-Ion Battery	-5°C to 40°C
2	Electronics	0°C to 40°C
3	Optical Sensors	20°C to 30°C
4	Solar Arrays	-150°C to 100°C
5	RF Antenna	-40°C to 60°C

5.1 SATELLITE WORKING ENVIRONMENT

A satellite has to work in different environments/phases viz. (i) Laboratory/ground environment; (ii) Launchpad; and (iii) In-orbit phase.

Heat transfer in a laboratory or ground environment is by conduction, convection and radiation. In view of the convective air medium and environment, with ambient temperature being around 25 degrees Centigrade, all the sub-systems housed inside the satellite will experience near-standard room temperature on the ground.

In the launch pad, heat transfer in a satellite is again by conduction, convection and radiation but the ambient temperature may vary depending upon the location of the launch site, season and time of the day. Hence the satellite may have to be heated or cooled by an external source to maintain the temperature of sub-systems within the set limits. For example, in Baikanur (Russia) Cosmodrome, the temperature may be as low as – 20 degrees Centigrade in winter whereas in Sriharikota it may be as high as 42 degrees Centigrade in summer.

During the launch phase, as the launch vehicle moves through the atmosphere, the satellite will experience rapid changes in pressure and temperature during the flight period. As the launch vehicle moves at very high speeds, the friction between air molecules and launch vehicle will result in very high temperatures. In this phase, the heat shield on the launch vehicle absorbs this energy and radiates to the environment thus protecting the satellite. Once the heat shield separates, the satellite is likely to get heated by the free molecules, which will be taken care of by the thermal design.

During the orbit phase, the satellite will be moving around the earth at an altitude higher than 200 kms and due to very low pressure, there will be no convective heat transfer and the heat transfer will be governed only by conduction and radiation.

Depending on its attitude, a part of the satellite will be exposed to cold space (typically ~ 4 degrees Kelvin) and part of it will be exposed to the sun and the satellite will experience extreme cold or hot temperatures in the range of -200 degrees Centigrade to +150 degrees Centigrade. The satellite will also pass through the eclipse (earth's shadow) and emerge again in the sunlight depending on the orbit and the type of the mission.

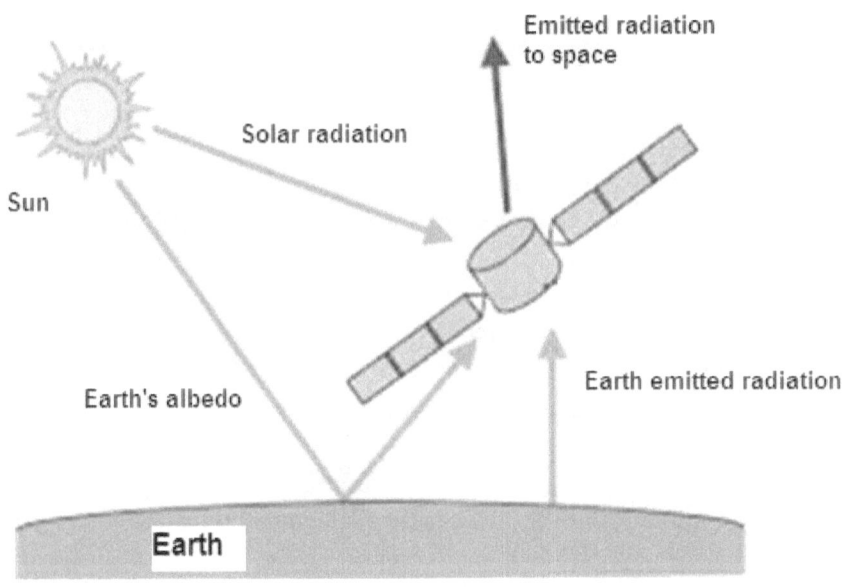

▲ **Fig. 5.1:** Thermal energy interactions with a satellite in orbit
(Courtesy: https://www.sciencedirect.com)

Fig. 5.1 shows various thermal energy interactions for a satellite in orbit. The heat input to a satellite is from the sun and the earth. The earth receives the sun's radiation, absorbs part of that radiation and reflects the remaining portion, which is called albedo and this will be around 35 per cent of the sun's radiation. Because of the earth's thermal equilibrium, the absorbed energy is radiated back to space which is termed as the earthshine. As the heat input to a satellite and heat loss in space are by radiation, one has to consider optical properties of surfaces used in the satellite and the wavelength of the radiation.

The emission or absorption of a surface depends on the wavelength and its properties are to be measured over these wavelengths. It may be noted that 98 per cent of the thermal energy in the sun's radiation is in the wavelength range of 0.2 to 3 microns and the earth's radiation is in the wavelength range of 5 to 50 microns.

Implementation of TCS in a Nanosatellite consists of three stages:
- Design and analysis
- Implementation of TCS elements on the satellite
- Testing the satellite for verification of TCS

5.2 DESIGN AND ANALYSIS

The design of TCS starts with the theoretical modelling of a Nanosatellite including factors such as:
- Surface area available for heat exchange
- Emissivity/absorptivity properties of materials used for the construction of satellite
- Thermal dissipation of sub-systems and their location on the satellite
- Identifying high power dissipating devices (hotspots)
- View factors between sub-systems
- Conductive and radiative paths available

Thermal analysis is carried out on the theoretical model by applying proper boundary conditions. For this, the satellite is divided into a number of nodes (using finite element model) and computing the estimated temperature excursions at each node. Thermal energy-balance analysis between the satellite and the space environment will be used to determine if the satellite has enough surface area to maintain specified temperature limits. The output of this analysis will be used in fine trimming the design aspects in an iterative manner. It is customary to take a conservative margin of about 10 per cent between the temperatures estimated by design with respect to the actual observed values.

5.3 IMPLEMENTATION OF TCS

After completing the design and analysis, the next phase is the implementation of TCS elements on the Nanosatellite which is explained below:

5.3.1 Types of thermal control

There are two types of TCS in a satellite - passive and active. Passive thermal control is achieved by using different materials with varied thermal properties to manipulate the heat transfer by conduction and radiation in orbit. Active thermal system is implemented using active elements such as heaters, which are operated with electrical power in a duty cycle mode. Passive thermal control is lighter and cost-effective than an active thermal control system. Passive thermal control defines any method of thermal management, which does not require energy input to function.

5.3.1.1 Passive thermal control

The following are the passive thermal control techniques commonly used in a Nanosatellite:
- Solid radiators
- Thermal blankets (MLI)
- Optical solar reflector
- Secondary surface mirrors
- White thermal paint
- Black thermal paint
- Passive radiators

a) **Solid Radiators**
Solid radiators or doublers are solid aluminium plates that provide large equipment mounting footprint for improved thermal dissipation through conduction. High power equipment is mounted using doublers instead of mounting directly on the panels.

b) **Multi-Layer Insulation (MLI)**
Multi-layer insulation (MLI) blanket is the most common passive thermal control element used on a satellite. MLI prevents both heat lost to the environment and excessive heating from the environment. All external surfaces of the satellite are covered in MLI blankets to maintain ideal operating temperature.

Multilayer insulation blanket consists of alternate layers of highly reflective shields separated by low conducting nylon net. An MLI blanket is fabricated with a number of layers of mylar or kapton sheets coated with a metallic material to reduce the radiation and separated by sheets of spacer material such as nylon net to avoid direct contact between adjacent foils. The external foil is usually single-side aluminised kapton facing out when exposed to space and is coated with gold to protect from UV and X-rays. For blankets used inside, the external foil coating depends on the application. Usually, double side aluminised mylar or kapton is used. **Fig. 5.2** shows a schematic construction of MLI blanket and **Fig. 5.3** shows a close-up view of a real MLI blanket.

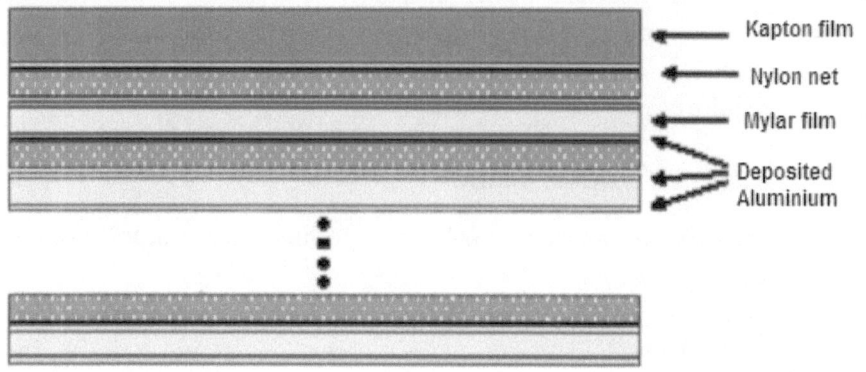

▲ **Fig. 5.2:** Schematic of an MLI blanket construction
(Courtesy: http://www.thermalengineer.com)

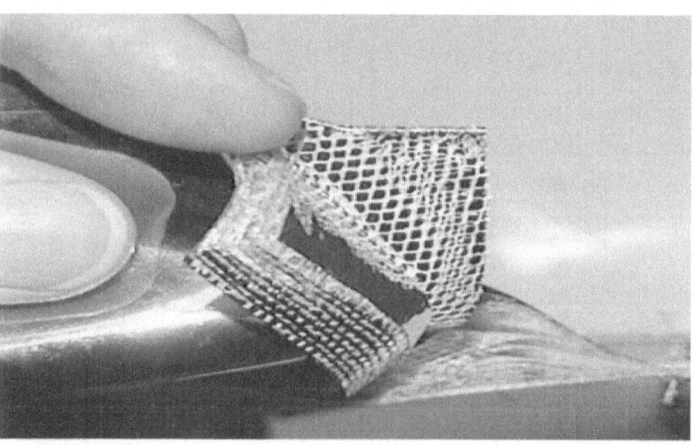

▲ **Fig. 5.3:** Close up view of a real MLI Blanket (Courtesy: https://en.wikipedia.org)

Because of the gold-coated outer layer of the kapton MLI blanket, most of the satellites in launch configuration appear in gold colour as shown in **Fig. 5.4**. The sheets are perforated to allow air passage during launch.

▲ **Fig. 5.4:** INS 1C covered with a thermal blanket (Courtesy: ISRO)

Factors affecting the efficiency of the blankets
The efficiency of MLI depends on the number of layers, type of coatings, the temperature of the layers and the pressure between them and most importantly the way in which the blanket is mounted. A single piece of blanket covering a large surface is more efficient than a number of small blankets covering the same surface. A blanket suspended over a surface is more efficient than one in direct contact with the surface.

c) **Optical solar reflector (OSR)**
The optical solar reflector is basically a mirror of the second surface with low absorptivity/emissivity ratio and negligible degradation in the space environment, which makes it an excellent cover for thermal control. It works like a radiator and is used to cover particular areas of the external surfaces of a satellite in order to reject undesirable heat into deep space.

OSR is made of fused silica glass with silver coating on the back surface to have low absorptivity coefficient α and high emissivity coefficient ε suitable for dissipating heat from components/equipment, maintaining lower temperatures for better performance such as amplifiers and batteries. **Fig. 5.5** shows typical OSR tiles mounted on a panel.

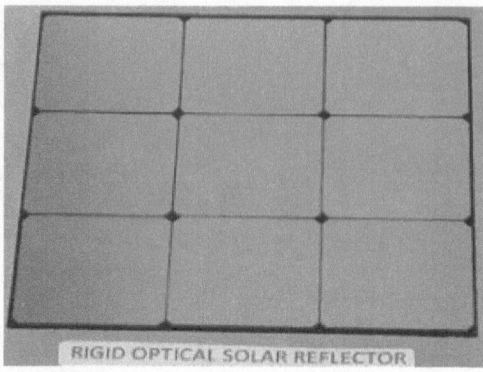

RIGID OPTICAL SOLAR REFLECTOR

▲ **Fig. 5.5. Rigid OSR mounted on the surface (Courtesy:** ISRO)

d) Second surface mirrors

Second surface mirrors resemble OSR in appearance and use. They are made of flexible plastic sheets instead of rigid glass. They are easier to handle but degrade faster with age due to charged particle bombardment and ultraviolet radiation (increase in α). These are used where rigid OSRs. cannot be used due to smaller surface area or curved surfaces. **Fig. 5.6** shows the schematic of a Second surface mirror construction.

Surface Exposed to Space

Conductive layer: Indium Tin Oxide (ITO)
UV Protection layers: C_eO_2
Base Film: Polyetherimide
Mirror: A_g
Corrosion Protection Layer: N_i alloy

▲ **Fig. 5.6:** Cross-sectional view of second surface mirrors

e) Surface coatings

Coatings are the simplest and least expensive of the thermal control techniques. A coating may be paint or a more sophisticated chemical applied to surfaces of the satellite to lower or raise heat transfer. The characteristics of the type of coating depend on their absorptivity, emissivity, transparency and reflectivity. The main disadvantage of a coating is that it degrades quickly due to the operating environment.

Black paint is commonly used to cover many electronic packages housed inside the satellite. The high values of α and ε of black paint help in maximising the heat exchange with other equipment. White paint is used on antennae to reduce temperature fluctuation and mechanical distortions to get better performance. **Fig. 5.7** shows an electronics package coated with black paint and **Fig. 5.8** shows antenna coated with white paint.

▲ **Fig. 5.7:** Typical electronic package coated with black paint (Courtesy: ISRO)

▲ **Fig. 5.8:** Satellite antennae coated with white paint (Courtesy: ISRO)

Passive Radiators: Excess heat dissipated on the satellite is rejected in space by using radiators. Radiators come in several different forms, such as spacecraft structural panels, flat-plate radiators mounted on the side of the spacecraft and panels deployed after the spacecraft is in orbit. Whatever the configuration, all radiators reject heat by infrared (IR) radiation from their surfaces. The radiating power depends on the surface emittance and temperature. The radiator must reject both the spacecraft's excess heat and any radiant-heat loads from the environment – hence generally they are mounted on the anti-sun side of a satellite. Most radiators are therefore given surface finishes with high IR emittance to maximise heat rejection and low solar absorption to limit the heat from the sun. Most spacecraft radiators reject between 100 and 350 W of internally generated heat per square metre. The weight of a radiator typically varies from almost nothing, if an existing structural panel is used as a radiator, to around 12 kg/m² for a heavy deployable radiator and its support structure. Deployable radiators on the International Space Station are shown in **Fig. 5.9**.

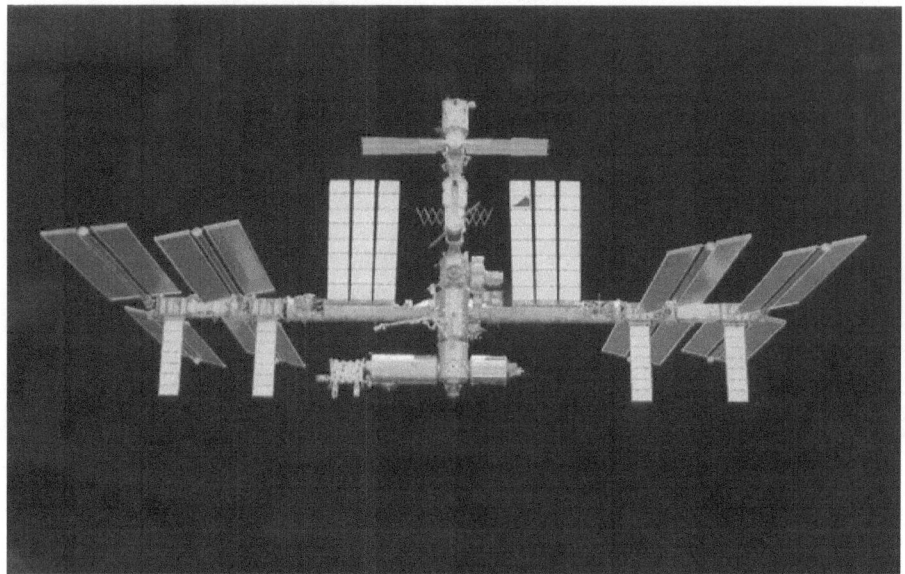

▲ **Fig. 5.9:** Radiator panels of International Space Station (in white colour)
(Courtesy: https://www.nasa.gov/)

5.3.1.2 Active Thermal Control

Some of the commonly used active thermal control elements are:

- Heaters
- Louvres
- Heat pipes

a) **Heaters**

Heaters are used in satellite thermal control to protect systems under cold environmental conditions or to make up for the heat that is not dissipated. Heaters are used with thermostatic or solid-state controllers to provide exact temperature control of a particular component. Heaters can also be used to increase the temperature of the components to their minimum operating temperature before the systems are turned on.

The most common type of heater used on a spacecraft is the 'thermo-foil' patch heater, which consists of an electrical-resistance element sandwiched between two sheets of a flexible electrically insulating material such as kapton. The patch heater may contain either a single circuit or multiple circuits depending on whether or not redundancy is required within it. Another commonly used heater is tape heater which is available as a long strip useful for wrapping around long components such as propellant plumb lines, radiator plates and optical sensor baffles. **Fig. 5.10** shows examples of both types of heaters.

(a)   (b)

▲ **Fig. 5.10(a):** Thermo-foil heaters (b) Tape heater on optics baffle
(Courtesy: https://www.minco.com/)

b) Louvres

Louvres are active thermal control elements used in many different forms. Most commonly they are placed over external radiators. Louvres can also be used to control heat transfer between internal spacecraft surfaces or placed on openings on the spacecraft walls. A louvre in its fully open state can reject six times as much heat as it does when it is fully closed with no power required to operate it. The most commonly used louvre is the bi-metallic, spring-actuated, rectangular blade louvre also known as Venetian-blind louvre shown in **Fig. 5.11**. Louvre radiator assemblies consist of five main elements - base plate, blades, actuators, sensing elements and structural elements.

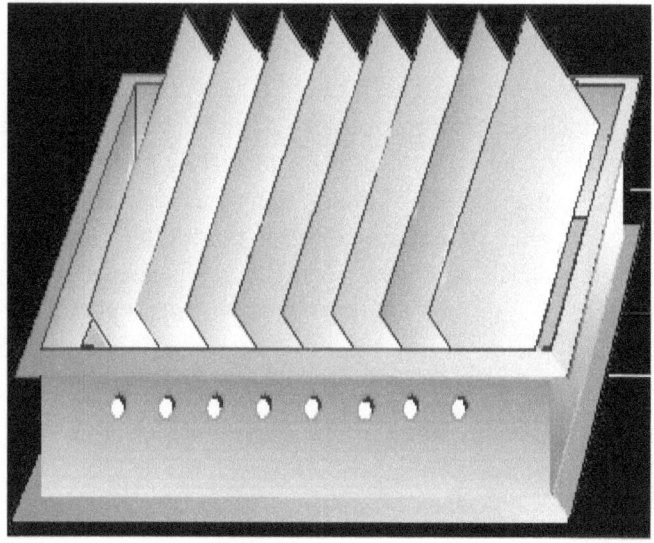

▲ **Fig. 5.11:** Venetian blind louvre (Courtesy: http://www.tak2000.com/)

c) Heat Pipes

A heat pipe is a device that absorbs heat from a hot spot and transfers it to a cold sink or a cold spot. A heat pipe consists of an envelope, a wick and a fluid. The fluid evaporates at the hot end and the vapour flows to the cold side where it is condensed to become fluid. The cold fluid returns to the hot end due to the capillary pumping action of the wick as shown in **Fig. 5.12.**

A network of heat pipes embedded in the honeycomb sandwich panel is called a heat pipe radiator panel on which equipment is mounted. This helps reduce thermal gradients in the panel.

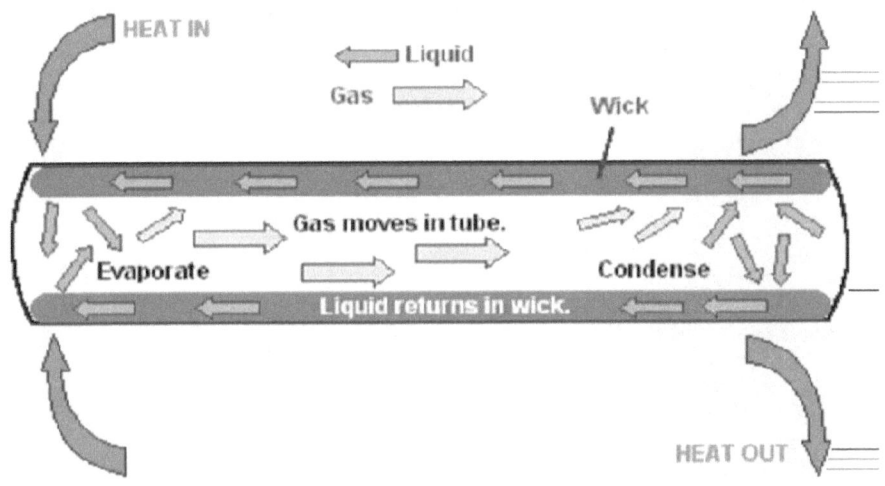

▲ Fig. 5.12: Working principle of a heat pipe
(Courtesy: https://www.yourdictionary.com/heat-pipe)

5.4 TCS IMPLEMENTATION IN A NANOSATELLITE

Thermal design of a small satellite differs from that of a traditional large satellite in a number of ways. Small satellites usually have much stricter mass, volume and power constraints for all sub-systems, including the thermal control system. Due to their lower mass, small satellites also have lower thermal inertia or capacitance which can result in more extreme temperature swings over an orbit.

In general passive thermal control techniques are employed in a Nanosatellite, which include:
- High emittance paint on the structure
- Heat sink for high power dissipation electronic devices (called "hot spots")
- Conductive and radiative couplings to transfer heat
- Insulating outer surfaces with MLI blanket with suitable cutouts for solar cells and sensors.

Active thermal control implemented in a Nanosatellite mainly consists of low power foil heaters to maintain the temperature of the satellite in case of any contingency. These heaters may be operated by ground command or with onboard auto logic. The thermal control design implemented in different Indian Nanosatellites is given below:

TCS of Nanosatellites designed by Indian students and launched by ISRO:

Pratham - Designed by IIT, Bombay, 10 Kgs: Passive thermal control. Heater for maintaining battery temperature

PI SAT - Designed by PES University, Bangalore, 5.25 Kgs: Active Control. Heaters for payload and battery

Swayam - Designed by College of Engineering, Pune, 1 Kg: Passive control. Kapton, white paint, low emittance tape and Optical solar reflector

Sathyabama
- Designed by
Sathyabama University,
Chennai, 1.5 Kg: Passive
control

Jugnu - Designed
by IIT, Kanpur, 3 Kgs:
Passive control. MLI,
OSR and surface
coating

SRMSAT - Designed by SRM University, Chennai: 10.9 Kgs: Passive control. MLI.

5.5   TESTING OF NANOSATELLITE FOR VERIFICATION OF TCS

It is important to verify the adequacy of TCS implemented on a Nanosatellite by testing under simulated space environment. Large satellites are tested in a solar simulation chamber using solar radiation simulator under vacuum but they are very expensive for a Nanosatellite. Hence Nanosatellites are subjected to a thermo-vacuum test in a suitable chamber simulating vacuum (10e-6 Torr) and external temperature conditions. All satellite bus systems and payloads are switched ON under both hot and cold conditions with temperatures of all sub-systems monitored using onboard and test specific temperature sensors. **Fig. 5.13** shows a Nanosatellite undergoing a thermo-vacuum test.

▲ **Fig. 5.13:** INS 1A under thermo-vacuum testing (Courtesy: ISRO)

REFERENCES

1. *'Spacecraft Thermal Control System'* by Col. John E. Keese
2. *'Thermal Control Sub-system for CubeSat in Low Earth Orbit'* by Harsh Vardhan Mishra, Department of Chemical Engineering, BITS, Pilani, India
3. *'Spacecraft Thermal Control'* by Prof. P. Rochus, ULG
4. *'Thermal Modelling of Nansat-Dai Dinh'*, San Jose State University
5. www.wikipedia.org

* * *

R. Arunachalam is a quality and reliability expert for the mechanical systems for various satellites of ISRO. His specialisation is quality assurance of satellite structures and mechanical elements, including deployment mechanisms and test evaluation of Satellite mechanical systems.

DEPLOYMENT MECHANISMS

R. Arunachalam

Early satellites belonged to the small satellite category and were limited by weight, volume and electrical power generation capabilities. As the size of satellites grew bigger so also the weight, volume and power requirements. Accommodating a bigger satellite on the launch vehicle was constrained by the volume available in the launch vehicle heat shield which is called satellite envelope.

Accordingly, the satellite design got modified with deployable solar panels, communication antenna and booms, which are stowed during the launch and deployed after the satellite is injected into the orbit, which has resulted in the evolution of deployment mechanisms. The reliability of deployment mechanisms is of paramount importance since any failure in this system can severely limit the useful life of a satellite.

With the advent of Nano and Microsatellites, which are small in size, weight and volume, that are launched on as piggyback with a bigger satellite, the constraint to accommodate the smaller satellites in the launch vehicle envelope is equally tricky. Generally, Nanosatellites require one or more mechanisms for deployment of solar panels and communication antenna, which may not be as complex as in larger satellites.

The reader will be happy to learn these aspects of deployment mechanisms in this chapter.

Large satellites usually have many flexible appendages such as solar panels, solar sail and boom, antenna reflectors and flap, which are stowed (due to volume constraints within the heat shield of the launch vehicle) on ground and deployed after the satellites are injected into the orbit. Hence a mechanical system (deployment mechanism) is required to deploy these elements. Deployment mechanisms are mission-critical and compactly stow appendages on to the spacecraft during the launch and deploy them positively once a satellite is injected into its orbit. The failure of the deployment mechanism will prove fatal to a mission and hence the mechanism demands very high standards of designing, manufacturing, assembling and testing.

Like a normal satellite, a Nanosatellite too can require one or more mission-critical mechanism for deployment of solar panels and communication antenna depending on the configuration. The satellite volume/dimension with deployable elements in stowed condition need to be compact to be accommodated within the volume available/provided by the launch vehicle (static and dynamic envelope of heat shield).

6.1 FUNCTIONS AND DESIGN CONSIDERATIONS OF DEPLOYMENT MECHANISMS

Generally, the deployment mechanism used in a satellite comprises two major elements:
• A hold-down and release device
• A mechanism for deployment

The hold down and release device ensures that the deployable appendages are rigidly fixed to the satellite's body by means of either tensioned wire rope or a tie rod to withstand launch loads during the launch phase and are subsequently released in orbit by means of an electro-mechanical pyro cutter to cut the cable/bolt to release the tension in the hold-down loop. The mechanisms provide the necessary energy/force for deployment. Most commonly used deployment devices are torsion springs and electric motors.

Deployment mechanisms have three principal functions:
• Deploy the sub-system in a smooth and controlled fashion, latching the sub-system at the end of deployment action with latch-up shock within acceptable limits/specifications.
• Provide the required stiffness for its final use/operation after deployment.
• Provide a delay/dampening mechanism to avoid locking shock and collision with the satellite body and its sub-systems.

The major environmental factors which influence the design of these mechanisms are mechanical vibrations, acoustically induced vibrations during launch, thermal gradients, vacuum and weightlessness in orbit. These are specific aspects irrespective of the type of satellite.

The hold-down mechanism must be capable of withstanding forces arising from vibrations in three orthogonal directions during lift-off and launch phase.

The in-orbit thermal environment (-150 degrees C to +100 degrees C) for mechanism elements could be very severe. Low temperature can lead to embrittlement of metals, weakening of adhesive bonding and increased friction in bearings. Large thermal gradients induce thermal distortion of parts/elements and may result in jamming of mechanisms. Hence one has to be very cautious in material selection for the mechanism system. The materials selected should have the capability to retain their properties within design limits over these temperature excursions. Particular mechanism elements namely bearings, limit switches and spring materials are susceptible to temperature excursions.

Under vacuum condition, loss of material can occur through evaporation or sublimation. All materials used in the mechanism must meet the specification of a total mass loss (TML) of less than 1% and collected volatile condensable matter (CVCM) of less than 0.1%. This environment also limits the use of conventional lubricants in bearings. The cold welding of molecularly similar metals in contact with each other is another possibility to be considered. In deployment mechanisms, dissimilar metals are used where there is metal to metal contact. Generally, such areas are dry lubricated with molybdenum disulfide. Some components are hard anodised for better wear resistance.

The design of the mechanism need not take into account the momentary load acting on the hinges during deployment of appendages. The deployable systems are provided with small actuating forces, which would be sufficient to provide the requisite energy for deployment in space.

6.2 CRITICAL ELEMENTS IN DEPLOYMENT MECHANISMS

The various critical elements in deployment mechanisms include:

a) **Springs:** Most commonly used element in a deployment mechanism which provides the motive force for different types of movements such as unfolding hinges, preloading the hold downs, locking of latches and kick off the movement in ejectors. Springs are basically termed as elastic structures. Beryllium copper alloy is an ideal spring material most commonly used as it is characterised by high strength, good electrical conductivity and good fatigue resistance.

b) **Latches:** They are used to deploy only devices. These are very critical components, since the system may not achieve full stiffness in the absence of a latch. The latches must be simple and designed with careful tolerance, especially to take care of large temperature excursions. The latches should have provision for minor adjustments such as length adjustment, which influence the final position of the appendage in orbit. Also, proper lubrication and wear issues need to be addressed.

c) **Dampers:** They are used to slow down the motion of the appendage being deployed. That is to say, they bring down the rate of deployment and latch-up kinetic energy of the system. Different types of dampers are eddy current damper and viscous damper.

d) **Sensors:** The common types of sensors mounted on deployment devices are limit switches to monitor the initiation or completion of deployment. Temperature sensors and accelerometers are other types of sensors used to get in-flight data.

e) **Dry Lubrication:** This is to prevent direct contact of two surfaces which are in relative motion. In the space environment, lubricants are essential to prevent damage from friction and wear. Solid lubricants applied on the components are typically for applications involving a medium number of duty cycles, moderate contact stresses and extreme environments. Essential properties that make a solid a good lubricant are low shear strength, good adhesion to surfaces to be lubricated and thermodynamic stability in the space environment. Most widely used solid lubricant in a satellite consists of films derived from MoS_2 (molybdenum disulphide) and materials such as poly tetra fluoro ethylene (PTFE). These are characterised by low friction coefficients in the range of 0.05 to 0.1.

f) **Pyro cutter:** This is an electro-mechanical device used to release appendages. It consists of a chisel edge cutter and an anvil. The wire rope to be cut is positioned in between these cutters and anvil. Initiated by an electric current pulse, the main charge gets ignited and imparts the motion to a piston which in turn imparts a linear motion to the cutter and the anvil. The cutter and anvil are mounted on two separate pistons and cut the wire rope. The pyro cutter, being a critical component, is equipped with both mechanical and electrical redundancy.

6.3 DIFFERENT TYPES OF DEPLOYMENT MECHANISMS

The following types of deployment elements are commonly used in Nanosatellites:
- Burn circuit
- Spring-loaded solar array
- Tape spring deployment
- Shape memory alloy (SMA)
- Storable tubular extendible member (STEM)
- Extendable booms

a) Burn wire mechanism for solar panel deployment
The deployment mechanism for the additional solar panels is based on a deployable solar panel hinge. Deployable solar panels are required when body-

mounted solar panels cannot generate enough power due to area constraints and/or the intended orientation (attitude) of the satellite in orbit needs deployed panels to generate adequate power. The solar panel deployment is activated electrically by using a burn circuit. The burn circuit breaks the thread that keeps the panel folded till the moment of deployment. The cutting of thread/cable releases the hinge-spring mechanism, effectively deploying the solar panel.

The release mechanism utilises a compressed spring system to ensure permanent contact with the cable that is securing a deployable panel. When heated the burn wire would cut through the cable due to a pre-compressed spring. After a successful cut of the cable the spring in the hinge extends to its natural length.

Usually, the burn wire is made of nichrome, which is heated to a high temperature and the cable is made of vectran (a high strength, high thermal stability and low creep fibre).

The advantage of this type of system is that it is small and relatively safer as compared to the pyrotechnic charge employed in bigger satellites. It, however, may fail to cut the wire in one-shot operation, which makes the system unreliable. To tackle this, usually redundant setups are used to ensure that the mechanism works reliably.

Fig. 6.1 is an example of deployment used in the Nanosatellite CAPE2, claimed to be the first 1U CubeSat to use deployable solar panels. The custom-built spring hinge and fishing line running through a resistance coil are used to deploy the panels.

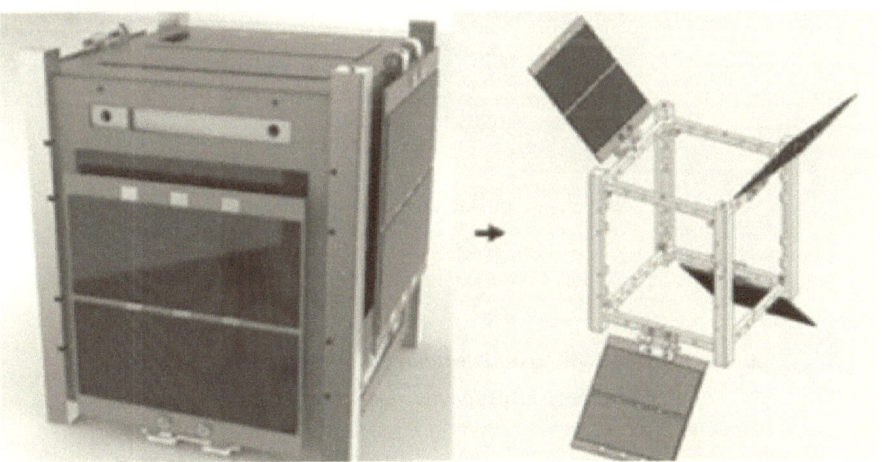

▲ **Fig. 6.1:** Burn wire mechanism (Courtesy: eoPortal Directory – CAPE2)

b) **Spring-loaded and tape spring deployment mechanism**

The spring-loaded hinge mechanisms are mainly categorised into two types: rigid type hinge and flexible type hinge. **Fig. 6.2** below shows both rigid type hinge and flexible type hinge.

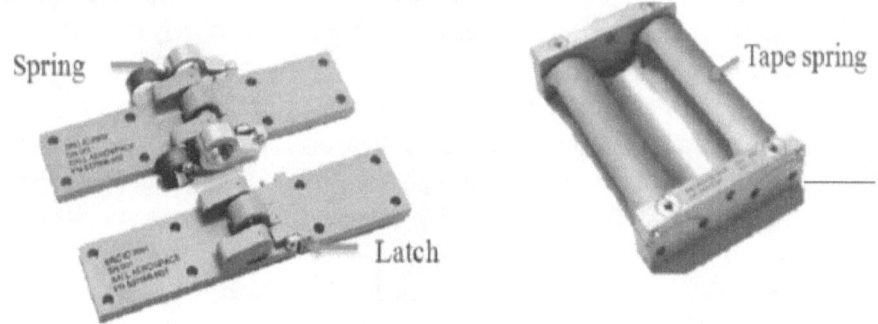

▲ **Fig. 6.2:** Types of spring-loaded hinges
(Courtesy: https://www.sciencedirect.com/science/article/pii/S2212667812000366)

Hinges of rigid type consist of a pin joint, two brackets, springs for deployment torque, latch and damper which can be added for low locking shock. The most common flexible type hinge is a tape spring hinge, which is also called 'lenticular' hinge or carpenter tape hinge.

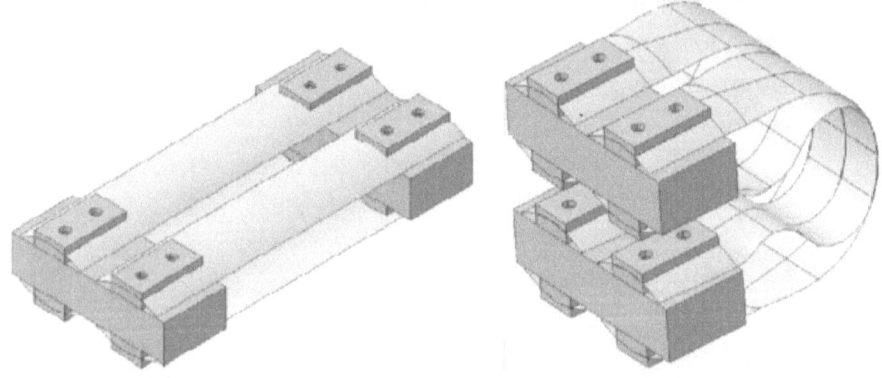

▲ **Fig. 6.3:** Tape spring flexible hinge before and after deployment
(Courtesy: https://www.researchgate.net/)

This type of hinge, shown in **Fig. 6.3**, consists of two brackets and tape springs, which are transversally curved strips. Although tape spring hinges have low bending stiffness when they are stowed for transportation (post-buckled state), they have relatively much higher bending stiffness due to the transversal curvature as tape springs become straight after deployment (pre-buckled state). In other words, a latch and an actuator for deployment torque are integrated into one element, a tape spring. Therefore the mechanisms of tape spring hinges are very simple and their mass is very low.

The rigid type hinges have good controllability of deployment behaviour and high stiffness but their cost is high. On the other hand, the tape spring hinges are a simple mechanism and have a lower cost. The performance of a tape spring mechanism is generally inferior to a rigid type hinge because it depends only on elasticity.

Fig. 6.3.1 shows different configurations of solar panel deployments of a typical CubeSat.

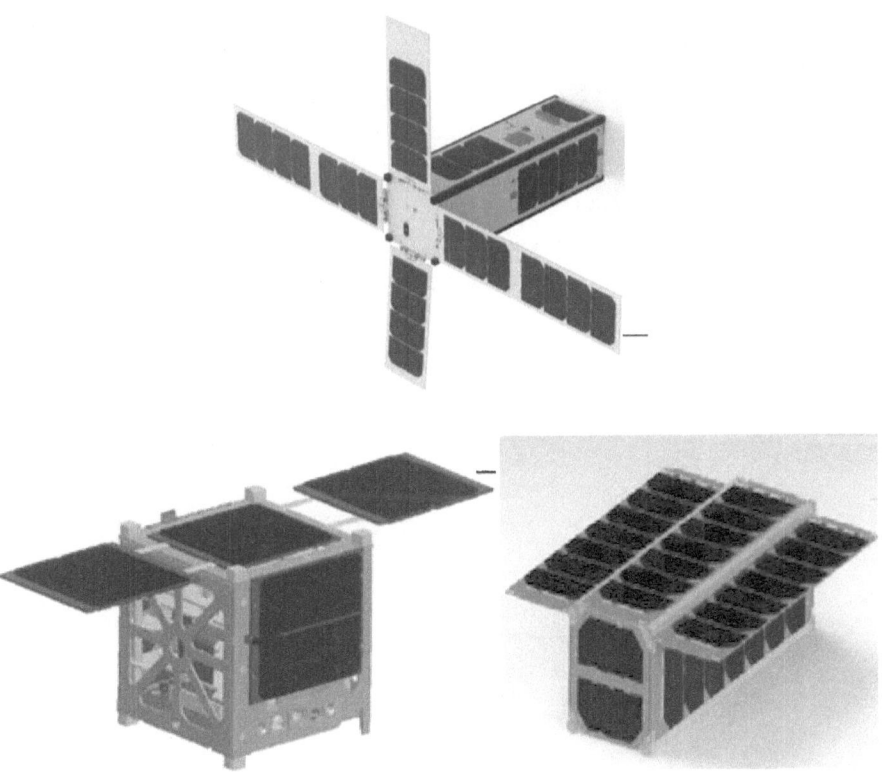

▲ **Fig. 6.3.1:** Different configurations of solar panel deployments
Courtesy: Gunter's Space Page

c) Shape memory alloy (SMA)

Shape memory alloys are metal alloys that "remember" original shapes and return to their original shape after being deformed by heating. They are also called smart materials. A mix of roughly 50% Nickel and 50% titanium called nitinol (NiTi) is the most common SMA. Also, copper-zinc-aluminium (CuZnAl) and copper-aluminium-nickel (CuAlNi) are widely used. SMAs have two stable phases - the high-temperature phase called Austenite and the low-temperature phase called Martensite. When the metal is changed to the Martensite phase simply by strain the metal becomes pliable and can withstand strains of up to eight per cent, which is called superelasticity.

Shape memory alloy hinge

A conductive hinge is made of a super-elastic shape memory alloy such as nitinol (NiTi) with a large elastic strain limit for enabling the hinge to bend around a small radius during stowage and flexible return to a trained rigid hinge position. These hinges conduct power through multiple solar cell panels that form a hinged solar cell array, resulting in the release of hinges from the bent stowed configuration to the rigid deployed configuration. The hinges further function as latches to lock the panels in place. A shape memory alloy hinge is shown in **Fig. 6.4**

The advantage of such hinges is that they are small in size and usually have a quick release time. They do not have pyrotechnic shocks. On the other hand, they may be susceptible to external rise in temperature and cause untimely deployment

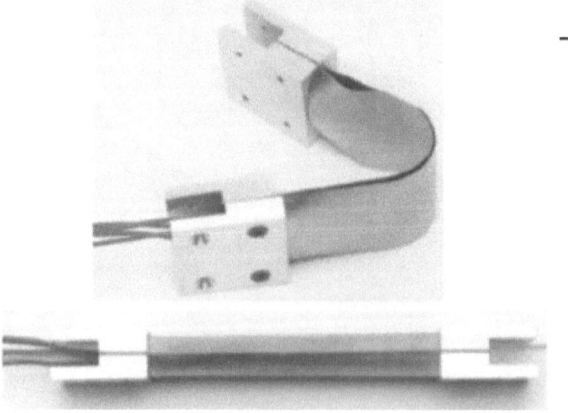

▲ **Fig. 6.4:** Shape memory alloy hinge in stowed (top) & deployed (bottom)
(Courtesy: http://citeseerx.ist.psu.edu/)

d) Storable tubular extendible member (STEM)

Storable extendible tubular member (STEM) is a deployable structural element. The STEM is a piece of steel or other material that rolls up flat on a drum that returns to its circular shape on deployment via a motor command. It is capable of pushing or pulling and is useful as a deployment structure for other booms. The stowed STEM fits into a small space and can extend many times its stowed length. A derivative of this is called the Bi-STEM, which can also be used as a deployable and retractable telescopic mast. The Bi- STEM has very high accurate positioning characteristics. A telescopic mast can employ a Bi-STEM (a two-piece STEM) boom as an actuator and stabiliser, which alleviates the need for the deployed telescopic mast segments to overlap. Due to this feature and because the segments can be fully overlapped when stowed, the mast enables an unusually lightweight and compact launch configuration. A typical CFRP STEM is shown in **Fig. 6.5** and configuration of STEM and Bi-STEM are shown in **Fig. 6.6.**

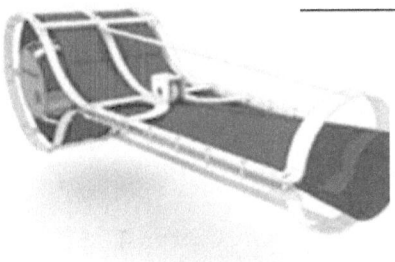

▲ **Fig. 6.5:** Deployable CFRP STEM and its mechanism
(Courtesy: https://www.researchgate.net/)

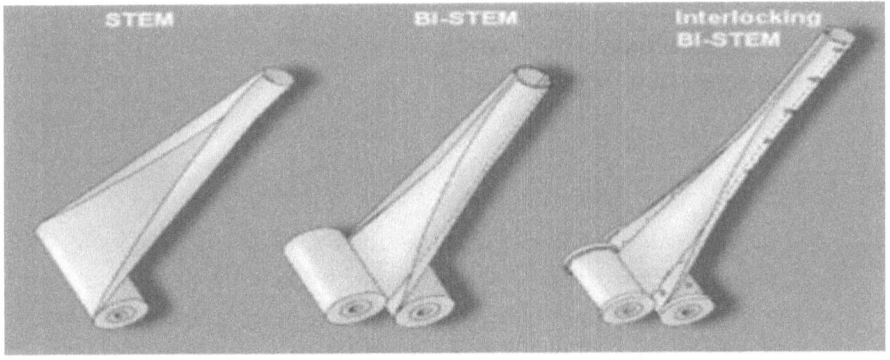

▲ **Fig. 6.6:** Different Configuration of STEM (Courtesy: https://ntrs.nasa.gov/)

e) Extendable boom

An extendable boom concept is used in the optics system to provide sufficient focal length for the imaging task. The extendible boom is deployed (after separation from the launch vehicle) with a lens attached at its tip as shown in **Fig. 6.7**. The deployment of the boom lengthens the distance between the lens and the imaging detector providing a long focal length. It implies that the boom structure must be aligned accurately and remain stable (thermally) under the various orbital lighting conditions.

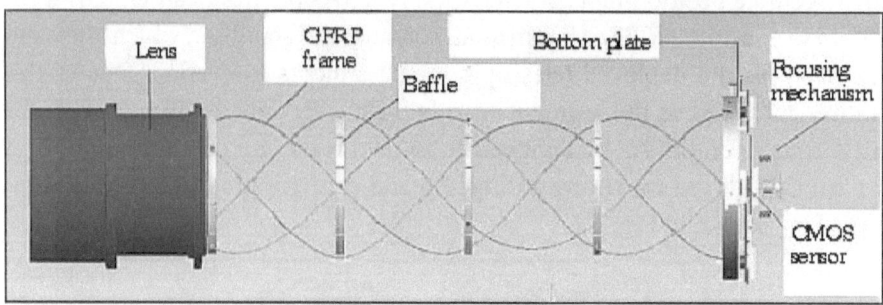

▲ **Fig. 6.7:** Deployment of lens to increase the focal length with extensible boom
(Courtesy: PRISM Nanosatellite.eoportal.org)

6.4 DESIGN CONSIDERATIONS FOR DEPLOYMENT MECHANISMS

Deployment mechanisms being mission critical, the system should have built-in redundancy with high force/torque margins. The design should preclude improper assembly and installation. The material selection should take into account the thermal environment as well as out-gassing, cold welding and lubrication. Adequate ground tests should be carried out to ensure successful deployment after exposing it to launch and on-orbit environments including mechanical, shock and thermal loads.

REFERENCES

1. 'Development of a Tape Spring Hinge with an SMA Latch for a Solar Array Deployment using the Independence Axiom' by Ju Won Jeong and Young Ik Yoo et al, Department of Mechanical Engineering, Korea Advanced Institute of Science and Technology, South Korea.

2. *'A Novel Tape Spring Hinge Mechanism for Quasi-Static Deployment of a Deployable Mechanism Using Shape Memory Alloy'* by Ju Won Jeong and Young Ik Yoo et al, Department of Mechanical Engineering, Korea Advanced Institute of Science and Technology, South Korea.
3. *'Space Flight 101 Space News and Beyond'*
4. Website: eoPortal Directory
5. *'Spacecraft Mechanisms – Spacecraft System Design Mae 342 Princeton University'* - Robert Stengel

* * *

R. Arunachalam is a quality and reliability expert for the mechanical systems for various satellites of ISRO. His specialisation is quality assurance of satellite structures and mechanical elements, including deployment mechanisms and test evaluation of Satellite mechanical systems.

07

POWER SYSTEMS

Ranganath Ekkundi

Electrical power is the lifeline of any space system whether it is a satellite, rocket or space probe. A satellite's power system is unique because the power is needed to be generated, regulated and distributed to various sub-systems continuously throughout the satellite's operating life.

In outer space, the sun is the main source of energy for power generation for all satellites that are orbiting around earth including Nanosatellites. Chemical batteries provide power to the satellite during the eclipse period and peak load requirements. The technology of solar cell and batteries evolved over decades based on power generation efficiency/capacity considerations.

The output from the solar arrays and batteries are available as DC voltage, which needs to be well regulated for usage by satellite bus and payload systems. Proper design of power flow path between these two sources and user loads ensures a smooth transition from sunlit to eclipse and vice versa, for charging of the batteries regularly. The desired DC voltage levels to the user systems are supplied equitably by power conditioning and distribution electronics.

The reader will get ample information about this branch of electrical and electronics engineering for Nanosatellite technology.

The main function of the power system in a satellite is to supply voltages to various sub-systems. This power system comprises three major sub-system activities viz., power generation, energy storage as well as power conditioning and distribution. Regardless of the size of a satellite, these are the essential sub-functions of a power system.

The electrical power needed for operating an in-orbit satellite has to be available from within. While it is possible for installing primary batteries for this purpose, they will only have a short life due to their limited capacity. For continuous availability of power over long durations, solar cells are used. In the outer space, the sun being the main source of energy, solar cells are the best option to convert sunlight into electricity by the photo-voltaic method. A number of solar cells are laid out and connected in panels to generate adequate power. Almost all satellites operating in orbit, small or big, use solar cells for power generation. Thus, Nanosatellites too are powered using solar cells.

Although sunlight is available most of the time in outer space near the earth, it so happens that an orbiting satellite traverses for some time through a region covered by the earth's shadow referred to as 'eclipse'. During such time, the sunlight will not be available and hence there will not be any power generation by the solar cells. To ensure uninterrupted power supply to the sub-systems of the satellite, a rechargeable battery is employed. Once the satellite crosses the shadow, the expended charge in the battery is replenished with the power generated by the solar cells. The use of a rechargeable battery serves a secondary purpose of meeting any peak requirement (such as payload operation), beyond the power generation capacity of the solar cells, for shorter durations even during sunlit periods.

There are, thus, two sources of power in a satellite viz., solar cells and rechargeable battery. It is important to understand that outputs from these two sources are available as DC supply and these outputs are not directly usable by the satellite as the DC voltage obtained is unregulated and is not of the desired value. In addition, control of the power flow path between the two sources and user loads is needed to ensure smooth switch-over during the normal sunlit and eclipse periods. This ensures the availability of desired voltage levels as well as proper replenishment of the battery. This function is carried out by power conditioning and distribution scheme embedded in the power electronics circuitry.

The three parts of the power system viz., solar power generation, storage in the rechargeable battery as well as power conditioning and distribution by the power electronics assure uninterrupted power supply to all electrical sub-systems of a satellite. While there are large communication satellites that demand 15 to 20 KW of power, a Nanosatellite typically consumes just 15 to 20 Watts.

The following sections describe these three aspects of the power system with special emphasis on how they are dealt with in small satellites.

7.1 POWER GENERATION

As explained above, power generation is accomplished by solar cells that convert sunlight into electrical energy. Solar cells are essentially semiconductor junctions that are exposed to sunlight wherein absorption of the photons results in the generation of charge carriers that easily cross the junction and become available to service any load connected at the output terminals.

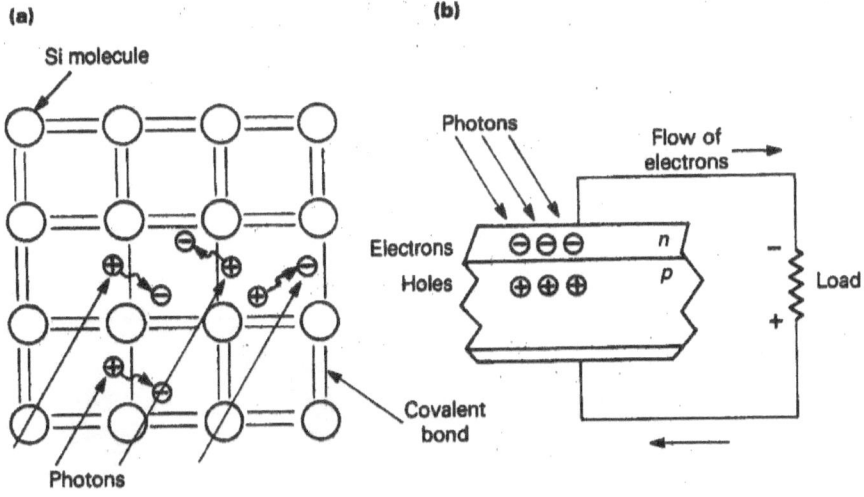

(a)

Si molecule

Photons

(b)

Photons

Flow of electrons

Electrons

Holes

n

p

Load

Covalent bond

▲ **Fig. 7.1:** Photovoltaic effect

Figure 7.1 above indicates this process as it happens in a silicon solar cell. One of the important characteristics of a solar cell is its efficiency. It is the ratio of power output by the cell to the power input as received by it from sunlight. Near earth, the average solar energy input is 1353 Watts per square metre (135.3mW per square centimetre). In the earlier days, silicon solar cells were configured with efficiency in the range of 10-12 per cent. Accordingly, a 2cmx2cm cell would provide a maximum output of 65 mW of power. How much of the incident energy gets converted into electrical output of the cell is dependent upon the spectral response of the material of the cell.

Subsequently, a lot of research and development (R&D) efforts have been made to identify new materials and processes to develop solar cells of higher efficiency. The development of gallium arsenide (GaAs) cells, as well as dual and triple junction cells, have paved the way to achieve ~30 per cent efficiency. Presently many advanced triple-junction solar cells are available in 4 cmx6 cm size, which provides nearly 1W output per cell.

Since sunlight has to penetrate into the region of the semiconductor junction, the solar cells are basically very thin. Typically, the thickness is in the range of 100 to 150 microns with the junction being just about 5 microns below the surface. The cells are also provided with a Cerium-doped cover glass of ~100 microns thickness in order to reduce the amount of radiation received by the active junction.

Altogether, with a thickness of less than 300 microns, these delicate cells need to be handled with great care. In a satellite, these cells are systematically laid down using adhesives on a more robust substrate creating a solar panel. In large satellites, these panels are of deployable type whereas, in small satellites, they may form a part of the outer surface itself and are referred to as body-mounted panels. **Fig. 7.2** shows a typical solar cell and a solar panel and Nanosatellites with deployed panels and body-mounted solar cells are shown in **Fig. 7.3**.

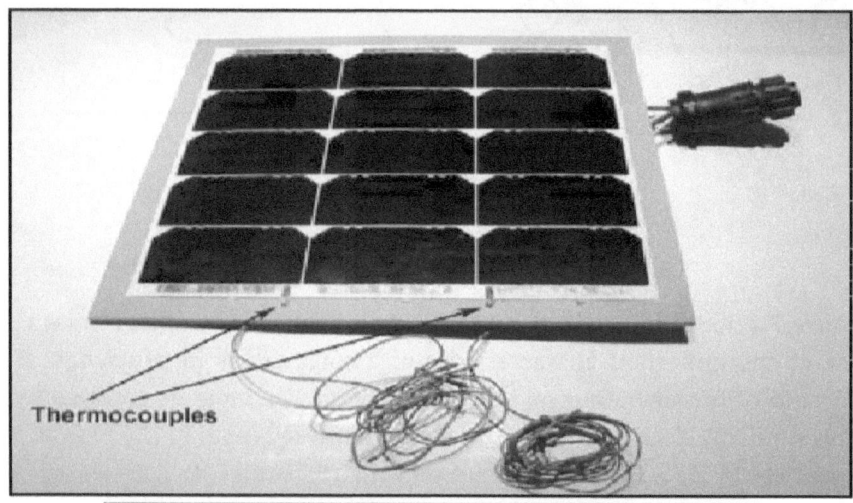

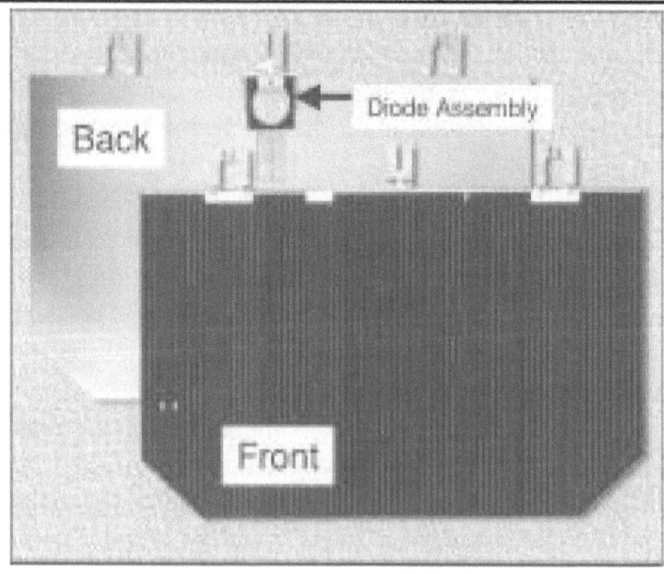

▲ **Fig. 7.2:** A typical Solar cell & Solar panel

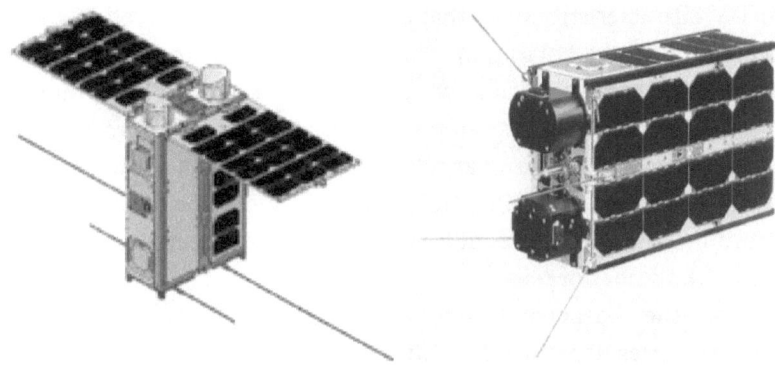

(a) Courtesy: Velox-II eoPortal   (b) Courtesy: Satnews

▲ **Fig. 7.3:** (a) Deployed solar array (b) Body mounted solar panels

Before starting the design of a power generation scheme with solar cells, it is important to learn about their power output characteristics and the dependency on various factors such as illumination, temperature and life span. **Fig. 7.4** shows the output characteristics of a solar cell. The I-V curve shows the variation of output voltage with the output current drawn by the load. When no load is connected at the output, the generated voltage by the cell is known as V_{OC}, the open-circuit voltage. As more and more current is drawn from the output by connecting a load, the output voltage keeps on reducing as shown by the I-V curve. When the output is fully shorted, the maximum current is delivered and is known as I_{SC}, the short circuit current. Hence a solar cell is basically a current limited source.

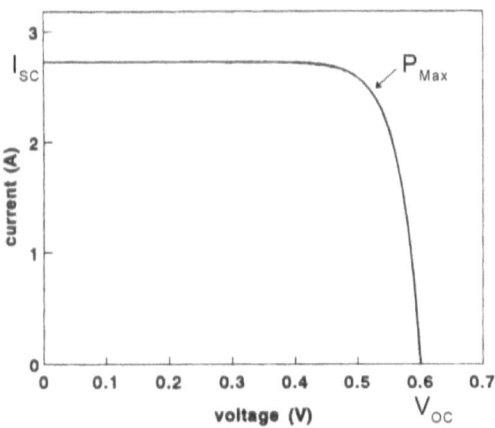

▲ **Fig. 7.4:** Output characteristics of solar cell

The I-V characteristics show that the output power delivered by the solar cell (which is the product of its output voltage and output current at any point), reaches a peak value at a particular point on the curve, known as the maximum power point or P_{mp}. The corresponding voltage and current values are referred to as V_{mp} & I_{mp}. By operating a solar cell at this point, we can extract maximum power from it. Hence one of the design considerations for the power system is to ensure that solar cells function at close to their maximum power point, optimising the number of cells.

However, the characteristics shown in **Fig. 7.4** are not fixed always as the I-V curve varies depending on three major factors viz., Illumination on the cell, temperature of the cell and the effect of particle radiation on the cell.

The sun's illumination intensity in the outer space near earth is referred to as AM0 intensity and its average value is taken as 1353W per square metre (the sun's intensity on the surface of earth at sea level is known as AM1 intensity and its average value is 1000W per square metre). The sun's intensity near earth is not always the same throughout the year. As the sun-earth distance varies the illumination also varies. While at equinox, the power incidence is 1353 W per square metre, during winter solstice it is 1399 W per square metre and during summer solstice it is 1305 W per square metre. These values are applicable when the solar panel faces the sun directly. If there is an off-normal orientation of any panel, then the incident power or intensity reduces as per the cosine factor. In a solar cell, the generated current is directly proportional to intensity whereas its voltage varies logarithmically with intensity. The effect of intensity on the I-V curve is shown in **Fig. 7.5**.

The operating temperature of cells also casts a major influence on the cells' characteristics. In orbit, depending on the season, satellite orientation and power drawn from the solar panel, the operating temperature of a solar panel and cell varies. While in eclipse a panel temperature could be as low as -100° Centigrade and in the sunlit part of the orbit it can reach up to +100°Centigrade. With the increase in temperature, the I_{sc} of the cell increases whereas its V_{oc} decreases. The effect of temperature on the I-V curve is shown in **Fig. 7.6**

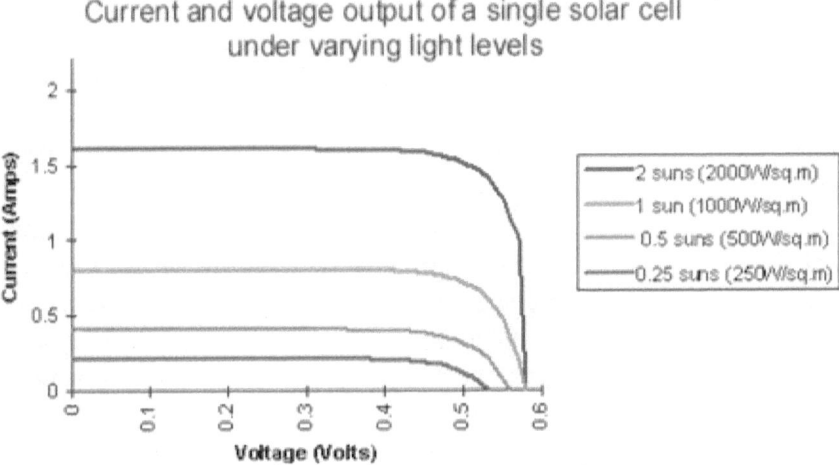

▲ **Fig. 7.5:** Effect of illumination on I-V characteristics (Courtesy: https://slideplayer.com/)

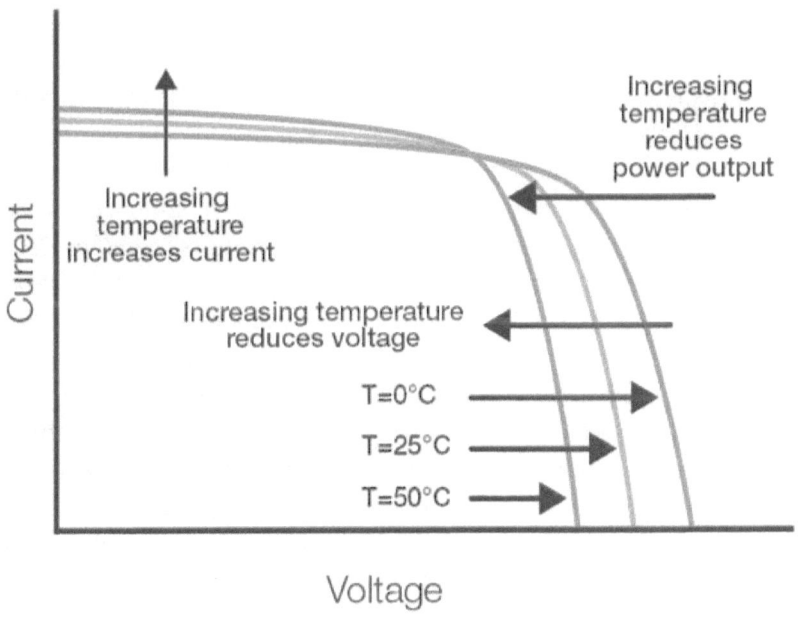

▲ **Fig. 7.6:** Effect of temperature on I-V characteristics (Courtesy: Seaward.com)

When the satellite is orbiting in space, it is exposed to radiation consisting of high energy particles. When electrons and protons impact an exposed solar cell, they create imperfections in the crystal structure of the cell thereby causing a loss in output power. This, termed as solar cell degradation, is dependent upon the accumulated dose of the radiation which in turn points to the lifetime spent in space. All solar cells are provided with cover glass to essentially reduce the amount of radiation dose received by the core cell. All manufacturers characterise their cells for radiation degradation and provide data. Based on the orbit and desired lifetime of a satellite, it is possible to estimate the total dose received using special model-based software programmes available for this purpose. Using this data in the designing of solar cell power generation, the output degradation is estimated.

Taking advantage of their high efficiency, all modern satellites, small or big, employ triple-junction solar cells in their design. These cells have proven heritage. There are at least four manufacturers of these cells, two in the US and two in Europe, who provide necessary data about their cells. The cells are normally available in sizes close to 4cmx6cm or 4cmx8cm. While designing small satellites, an important consideration is given to establish a good match between the size of the panel and the size of the cell. This is because one has to pack as many cells as possible in a given area to maximise the utilisation of available panel size.

Solar cells are manufactured in relatively smaller sizes. In earlier days the cells of 4 or 8 square centimetres were available. Present multi-junction cells are available in nearly 30 square centimetres size. Typically, one cell develops 2.5V and 400 mA of current. In a panel, a number of cells are serially connected to generate the desired output voltage and many such series are connected in parallel for generating the desired current. To interconnect the cells, a special type of 'silver interconnects' are employed. These have loops to take care of thermal expansions and contractions that are expected due to large temperature variations in the panel as a result of sunlit/eclipse transitions in orbit. These cells are laid out on a panel by a controlled process of applying adhesives. This whole process depends on the type of cell, the substrate material and the applicable environmental conditions. The process involves a concerted effort and is a technology in itself.

The panels need rigorous testing to ensure reliability. Electrical performance is evaluated under a sun simulator set up; the acoustics are tested so that the panels can withstand launch environment; thermal cycling tests check the integrity of the fabrication to withstand expected temperature variations. Since all these tests need special facilities, for small satellites, these can be part of procurement specifications for solar panels.

7.2 ENERGY STORAGE

Rechargeable batteries are used for energy storage in a satellite. The rechargeable batteries employ electrochemical cells that convert chemical energy into electrical energy. The internal chemical reaction being of reversible type, they facilitate the process of re-charging, unlike in primary batteries, wherein once discharged, those cells need to be physically replaced.

A battery cell typically consists of four parts viz., anode, cathode, electrolyte and a separator. The chemical materials used for these parts are different for different types of cells. For space applications, three different types of cells are used. They are nickel-cadmium, nickel-hydrogen and lithium-ion. The capacity of a cell is specified as ampere-hour (AH). It is the amount of current in amperes that a fully charged cell delivers in one hour till its voltage reaches down to its end of discharge value. A comparison of typical performance of these batteries is shown in **Table. 7.1**.

These batteries discharge during the eclipse phase and get recharged during the sunlit period. The pattern of occurrence of the eclipse is dependent on the orbit. In a typical LEO, there can be 36 minutes of eclipse period and 65 minutes of the sunlit period in one orbit. In such a scenario a fully charged battery, in one orbit undergoes one cycle of discharge and charge. Thus, the battery goes through 14 cycles per day and 5,110 cycles per year. While most of the small satellites are designed for one to two years of life, many remote sensing satellites are designed to work for five years in which case the number of cycles works out to be greater than 25,000. Communication satellites in GEO on the other hand experience just 90 cycles per year. This works out to be a comparatively meagre 1,350 cycles during the satellite's in-orbit life of 15 years. As a cell is charged and discharged many times, its performance degrades over time, because following each reversal of a reaction, the reactants do not revert fully and identically to the original composition and quantity.

The extent of degradation depends on the depth of discharge (DOD) that takes place in each cycle as shown in **Fig. 7.7**. The DOD is the ratio of the quantity of AH taken out in a cycle of discharge to the capacity of the cell. Lower the DOD, the higher is the number of the charge-discharge cycle that can be serviced within acceptable degradation. Hence generally, LEO satellites operate with a DOD of 20 per cent whereas GEO satellites work at a level of 70 to 80 per cent.

Advances in space battery technology are aimed at increasing the energy density and life cycle so that the same amount of energy support is given with lesser mass impact. In earlier days nickel-cadmium and nickel-hydrogen cells were being used. The present state-of-the-art cell is lithium-ion. It has several advantages over the other two and is extensively used in both large and small satellites. **Table. 7.1** summarises the comparison between these battery technologies.

▼ **Table. 7.1:** Comparison of different types of Battery

Parameter	NiCd	NiH2	Li-Ion
Energy density (Watt-hour/kg)	35	55	>100
Watt-hour/litre	80	60	>200
Cycle life	20,000@20%	40,000@20%	40,000@20%
Calendar life (years)	5-7	20	>15
Storage life (years)	>4	2-4	3-5
Cell voltage (V)	1.2	1.25	3.6
Temp. range(°C)	-5 - +40°C	-20 - +20°C	+ 10 - +30°C
Self discharge(%)/month	15-20	60	< 10

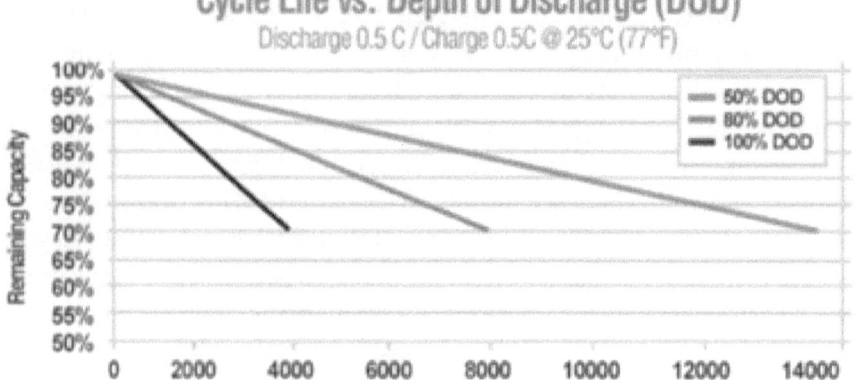

▲ **Fig. 7.7:** Effect of DOD on battery cycle life
(Courtesy: https://modernsurvivalblog.com/)

Presently Li-Ion cells are available as either low capacity COTS type or high capacity cells. Sony and LG cells have about 2.5 AH capacity whereas SAFT and MELCO have capacities of 45, 50 and 100 AH. The high capacity cells are used in high power big satellites and COTS cells are extensively used in Nanosatellites. Different types of Li-Ion battery cells are shown in **Fig. 7.8.**

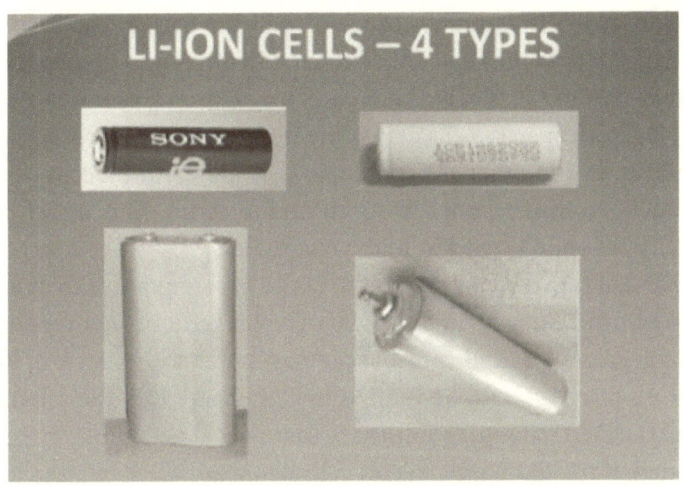

▲ **Fig. 7.8:** Types of Li-Ion battery cells

The operating voltage range of a Li-Ion cell is between 4V and 2.5V. In order to achieve the desired voltage from a battery, a number of cells are connected in series and to achieve the desired capacity, the cells have to be connected in parallel. Unlike Ni-Cd and Ni-H2 cells, Li-Ion cells can be easily connected in parallel. In a battery, the cells are connected in series and parallel.

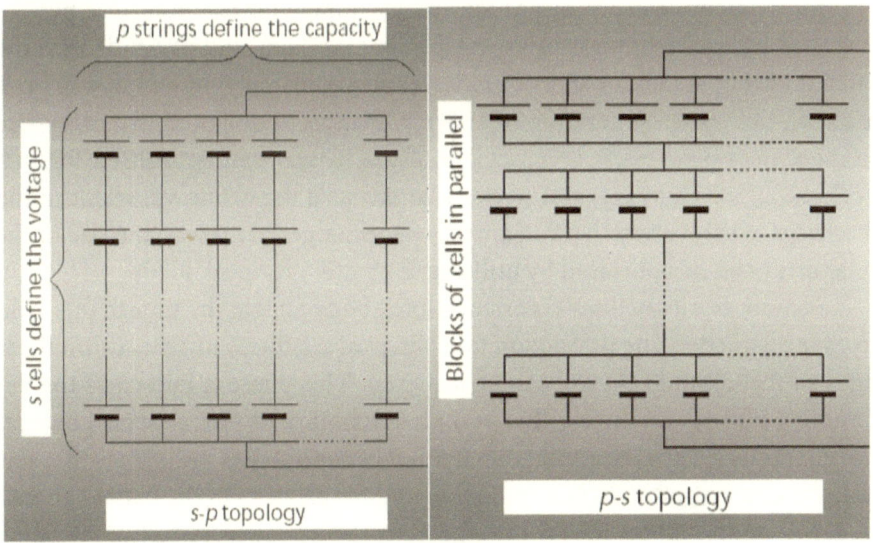

▲ **Fig. 7.9:** Two different ways of realising the Li-Ion battery

There are two different ways in which a Li-Ion battery can be utilised. As shown in **Fig. 7.9** cells may be connected first in series and then each series set can be connected in parallel at the ends. This is known as s-p topology. This is suitable for low capacity cells and is the most appropriate configuration for small satellites. In the other p-s topology, cells are connected in parallel first and then each paralleled bunch is connected in series. This is well suited for high capacity cells and is used in high power satellites.

Since Li-Ion cells are not tolerant to abuse, it is important to operate them with extreme care. They can neither be overcharged nor be over-discharged. This calls for protection and control schemes in the power system. Preferably, the rate of discharge should be limited to C/2, where C is the capacity of the battery and the DOD should be limited to the allowed maximum in the mission. When loaded, if the discharge voltage falls below 3V for a cell, the loads need to be disconnected. While charging, the rate of charge should match with the energy balance requirements such that the discharged power is replenished to the full extent. In order to avoid overcharging, an appropriate end-of-charge scheme should be chosen when the full charge voltage is reached. Normally, a scheme where the current is reduced in steps is best suited to ensure that the battery gets fully charged without experiencing overvoltage at its terminals. For the battery to perform at its best, its operating temperature range needs to be narrowed by thermal control.

Although a battery is meant for energy storage for use in eclipse period, it can be utilised for two more special purposes. First, the payloads in some satellites are operated for short durations and accordingly they appear as peak loads on the power-generating solar array. Instead of designing the array size to meet such peak loads, the services of the battery can be utilised for that peak duration. This helps not to oversize the solar array but will result in the discharge of the battery that can anyway be recharged. This way, the size of the solar array can be optimised by utilising the battery for peak loads.

Second, small satellites experience some body rates at the time of injection from the launcher. The first action for control of satellite is to bring down these rates so that attitude control can be achieved. This phase is known as the de-tumbling phase, and during this period, availability of proper solar power is not guaranteed since the attitude itself is not in control. It is the battery that acts as the source of power during this phase and the capacity of the battery should be sufficient to support the load till the de-tumbling is successfully completed. Sometimes the capacity of a battery is decided based on this requirement rather than by the regular eclipse energy requirement.

Fabrication of battery involves various steps. One of the important criteria is to match and select cells for their performance. Well-matched cells are used for parallel and series connections. Cell to cell interconnections are done through welding and cells are held tight together with robust mechanical design practises that involve bonding and securing through mechanical fasteners. The cells dissipate heat during their operation and an appropriate thermal path is also to be factored in the design of a battery. If needed, heaters are to be planned right as a part of the battery assembly. Connectors are used for interfacing with the satellite. The overall satellite design should facilitate charging the battery from an external source even after the satellite is fully assembled. In Nanosatellites, a battery is fully charged at launch base before integrating with the launch vehicle.

Batteries have to be tested for their charge and discharge characteristics before they are delivered for integration. Batteries are also characterised at an expected temperature range. In small satellites, the battery vibration test may be carried out along with satellite itself.

7.3 POWER CONDITIONING AND DISTRIBUTION

The major functions of power conditioning and distribution of the power systems are:
- DC bus formation
- Solar array power management
- Battery charging control
- Battery protection
- Power distribution

A power bus is a common point into which the power from the two sources - solar array and battery - is fed before distribution. Since the characteristics of both sources are different, a planned interconnecting of both to a common bus is necessary. Many different ways of bus formation are in practice.

An unregulated bus, as shown in **Fig. 7.10**, is a simple system wherein the solar panel output is tied to the battery. It is a highly efficient system in that it allows for direct energy transfer from both sources to the bus. The bus voltage varies with the battery voltage, which will decrease in the eclipse phase and increase in the sunlight phase. The shunt charge regulator protects the battery from overcharging.

In an unregulated bus, the solar array operates at varying voltages that follows the battery voltage. An improved version of this is the sunlit regulated bus in which the solar power is regulated during the sunlit phase to its maximum power point value. The battery is connected to the bus through a diode which stays reverse biased in the sunlit phase and becomes forward biased in the eclipse phase. During the eclipse phase, the bus voltage decreases as per the battery voltage. In large satellites where the solar panels are deployed and made to track the sun always, this scheme is useful since the maximum power point is fairly fixed. The sunlit phase regulated bus is effectively employed in GEO communication satellites, wherein the eclipse phase is just 90 days in a year.

A fully regulated bus has a fixed bus voltage during both the eclipse and sunlit phases. For this, a battery discharge regulator (BDR) is used as shown in **Fig. 7.11**. In this case, there will be a certain power loss in the BDR. This bus is useful for the user sub-system design since it always receives a constant voltage. This scheme is used in high power satellites such as communication satellites.

For small satellites, the unregulated bus is well suited since it supports direct energy transfer. However, the solar array output will be non-optimal as its operating voltage varies with the battery voltage. The amount of variation in battery voltage depends on the DOD. Since the DOD is kept generally small, this off optimal loss is also not significant.

An alternative scheme employing maximum power point tracking (MPPT) regulator is shown in **Fig. 7.12**. But this calls for extra hardware in terms of battery charge regulators (BCRs), which force a section of the solar array to operate at its maximum power point. A judicious choice has to be made between these two schemes.

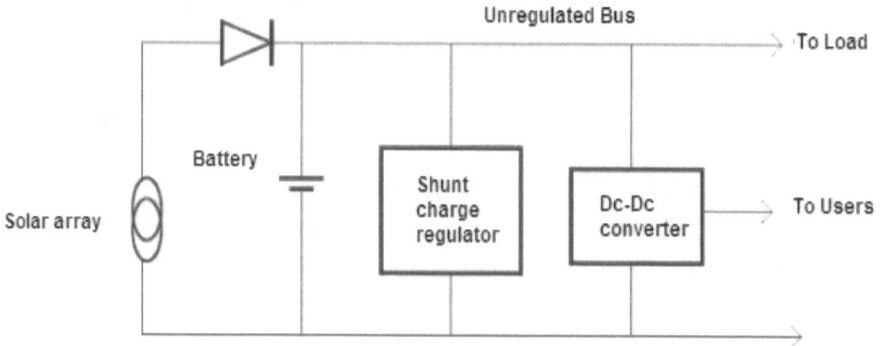

▲ **Fig. 7.10:** Schematic of unregulated bus

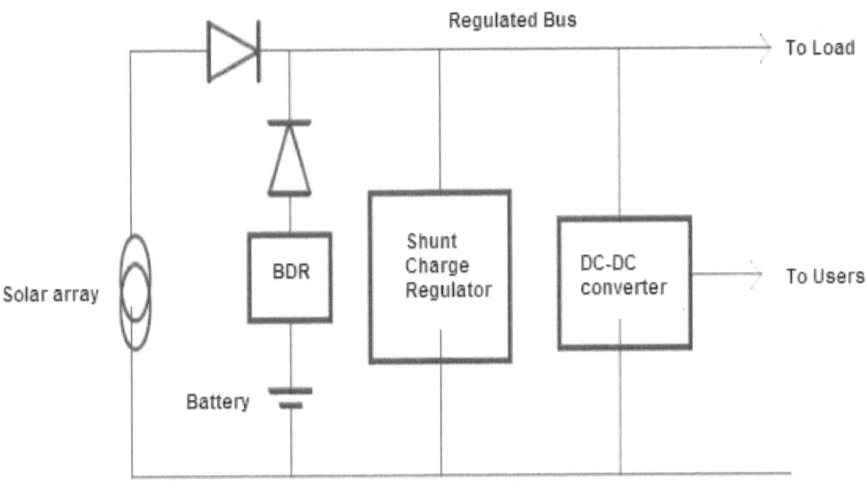

▲ **Fig. 7.11:** Fully regulated bus

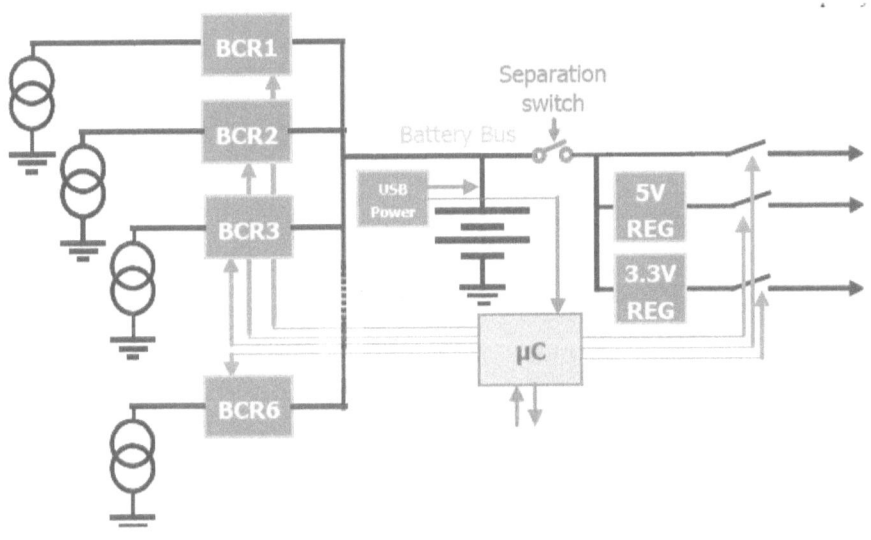

▲ **Fig. 7.12:** MPPT based power bus

For battery charging control, in a direct energy transfer bus, an LTP-UTP based control can be adopted as shown in **Fig. 7.13**. The principle of operation is as follows: Initially, the voltage across the battery is low and the shunt switches are open. The battery starts charging with a current of I source - I load. When the voltage across the battery reaches the UTP level, the shunt switches are on and the strings are shunted. Then, the 'BAT' starts discharging through the load. When the voltage reaches LTP it turns 'OFF' the switch and this process repeats.

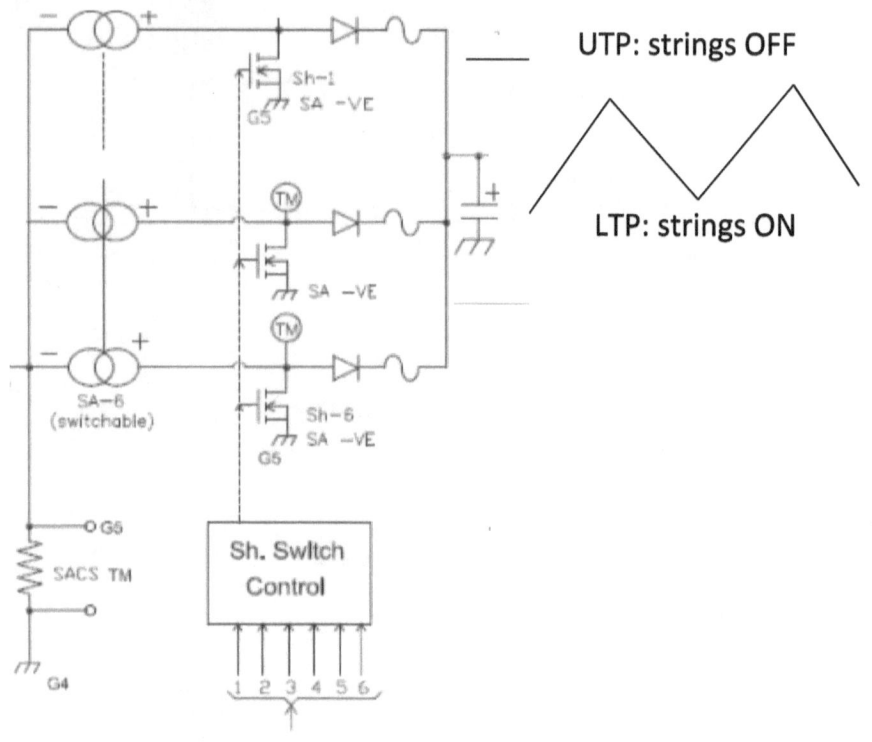

UTP: strings OFF

LTP: strings ON

▲ **Fig. 7.13:** LTP-UTP based Charge control

In a small satellite, protecting a battery against discharge can be implemented as a part of the onboard computer (OBC) operation. Here, if the battery voltage falls below a set value, the OBC can be programmed to send signals to switch off loads.

Power distribution is the connection between the bus and user subsystems. This is done through DC-DC converters, which receive the bus voltage as input and provide regulated voltages as the output as per user requirements. DC-DC converters work on the principle of pulse width modulation (PWM) control.

All the power conditioning and distribution schemes are configured in the form of electronic circuits. These are implemented in the standard way in which any electronics hardware is made.

The testing of these systems will be as per the QA norms of testing any electronic system. This will normally include vibration and thermo-vacuum tests.

7.4 POWER SYSTEM DESIGN

To design a power system for a small satellite, the inputs required are the orbit, sub-system load power requirements and the design life. From the orbit, one can derive information about the sunlit and the eclipse durations as well as their pattern. From the data of individual load power requirements, one can compute the total power required during the sunlit and the eclipse phases and also take note of any peak load and its duration. The design life information is to be used to consider solar array degradation and cycle requirements of a battery. The following steps give an outline of how power system elements can be designed.

The product of the eclipse duration and the power required in this phase determines the energy taken out from the battery. $W_h - = P_e \times T_e$

With this, we can compute the ampere-hour used from the battery by dividing the energy by the battery discharge voltage. $A_h - = (W_h -)/Vd$

Based on the cycle life required, the DOD can be estimated by referring to the battery's characteristics.

If the ampere-hour used from the battery is divided by this DOD, the capacity requirement of the battery can be easily arrived at. **Battery rating = A_h-/ DOD**

The energy required to be replenished in the sunlit phase can be computed as $W_h + = W_h - *1.1$

If this is divided by sunlit duration, then the power needed for charging the battery as P_{ch} can be arrived at.

Adding this to the load power will give the total solar array power that needs to be generated. Referring to the panel manufacturer's data, the adequate panel size can be planned.

It is to be noted here that the battery capacity as arrived at by the above calculation assumes only the eclipse load. If there is an advantage by planning peak load support from the battery, then accordingly the extra cycles and expected DODs will be taken into account in firming up the battery's capacity. Once the battery's capacity is chosen, the extent of battery support for the initial loads during the de-tumbling phase will be checked so that sufficient capacity from the control point of view can be ensured.

7.5 SPECIAL CONSIDERATIONS FOR NANOSATELLITES

Small satellites are launched as 'also flown' satellites in a launcher. This has two impacts. One is that the choice of orbit gets restricted since it may have to match with that of the major satellite on which the Nanosatellite is riding piggyback. This has to be considered in finalising the overall mission and schedule. Second, the launcher restricts small satellites to be kept completely in the 'OFF' mode during the launch. As a result, there is a need to design a scheme to turn 'ON' the power after the small satellite separates from the launcher. All initialisations have to take place after the Nanosatellite is powered on. The scheme to detect injection from the launcher has to be discussed with the launcher agency and accordingly the circuit for auto turn 'ON' of the Nanosatellite has to be designed.

As mentioned earlier, the battery capacity should be good enough to support the load during the duration of de-tumbling as expected from the control performance.

Small systems have to be of small size and lighter mass. Accordingly, multi-junction solar cells and Li-Ion batteries are the primary choices for the power system design. The solar panel size has to be well-matched with the solar cell size to facilitate dense packing of the cells in the given area. Even the power electronics hardware has to be compact. Towards this, ASICs and FPGAs are to be planned. The use of miniature DC-DC converters available in the market should be judiciously planned to meet different load requirements.

* * *

Ranganath Ekkundi is a satellite power systems expert and specialised in Power conditioning and distribution Electronics. He made significant contributions and innovative designs in the design and development of high power electronics for many Communication and Remote sensing satellites.

ONBOARD COMPUTER & DIGITAL ELECTRONICS

K. Parameswaran

Onboard computer (OBC) and digital electronics are the living link between the analogue and digital systems and are the brain of the satellite. These systems perform a host of satellite housekeeping functions, which cover the design of telemetry, telecommand, attitude control electronics, thermal management and payload operations.

Telemetry sub-system adopts the technology of digital encoding, encryption, processing and digital modulation techniques to provide satellite health parameter data through RF downlink from the satellite to the ground station. The telecommand system adopts the digital coding, error correction and digital demodulation techniques for generating and controlling satellite functions from the series of commands sent regularly from the ground station to the satellite.

Attitude control electronics (ACE) receives the spacecraft orientation details defined as attitude information from onboard sensors and processes the same in accordance with control algorithms for generating control signals for actuators to move the satellite to the desired orientation.

The onboard computer (OBC) is designed to perform all the above functions systematically through the computational power of the hardware and software onboard. The software follows modelling and elaborate software development methodology to realise the designed parameters with adequate margins.

The reader will get sufficient information on these technologies for the Nanosatellite in this chapter.

The onboard computer (OBC) is the brain of a Nanosatellite configured as an integrated electronic system where all functions such as telemetry (TM), telecommand (TC), attitude control electronics, thermal management and operation of the payload as well as all other sub-systems are implemented.

An OBC is configured to perform the following functions:
• Receive and respond to commands from the ground station

- Collect the satellite health data, format and encode the data for transmitting to the ground station
- Attitude determination and control by using appropriate control algorithms
- Monitor temperature of all sub-systems and maintain a specific temperature range by using heaters
- Store the TM data during non-visible periods of the orbit and download during radio visibility

A typical OBC block diagram is as shown in **Fig. 8.1**

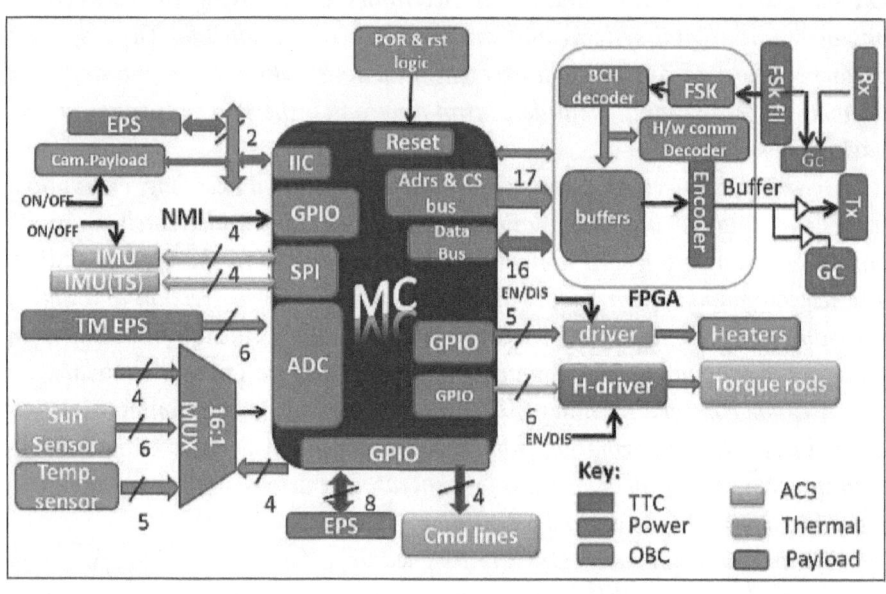

▲ **Fig. 8.1:** Typical OBC block diagram (Courtesy: eoPortal, PISAT)

8.1 OBC HARDWARE

An OBC for Nanosatellites can be configured as a firmware using micro-controller, microprocessor, FPGAs and other peripheral devices such as PROM, RAM and standard I/O interfaces. Different versions of a total OBC implemented on a printed circuit board are commercially available. A typical OBC design configuration and interfaces are explained below:

CPU
Types of CPUs normally used are:
Honeywell HX1750 radhard16 bit microprocessor

Radhard 32 bit SPARC V8 processor – UT699
Microcontroller available as IP core
Processor configured within an FPGA ported as IP core
Clock Speed: 12 MHz / 24 MHz

Memory
32K x 8 PROM
512K x 40 RAM
512K x 40 EEPROM

Digital logics
RTAX 2000 FPGA (ACTEL anti-fuse)

Interfaces
UART for GPS, reaction wheel
I2C for Payload
SPI for IMU
MIL-STD- 1553B

Radiation performance (optional)
Total dose of 100K rad (Si) according to MIL-STD-883 method 1019
SEU error rate better than 1E-5 errors/device/day
No single event latch ups below a LET threshold of 70Mev cm2/ mg

8.2   OBC SOFTWARE

An OBC software consists of several modular components such as control modes of operation, TM, TC and safety logics. Standard development tools and development environment are used. Safety logics consist of watchdog timer and remote programming.

8.2.1   Watchdog timer (WDT)

When a microprocessor is executing its programme normally, it resets the WDT once in every major cycle. If the microprocessor fails to do so, the WDT issues an interrupt to the microprocessor which in turn jumps to the programme reset point and retries the execution of the programme.

8.2.2 Remote programming

To take care of unforeseen disturbances encountered in an orbit, a feature called remote programming is incorporated in the OBC. Normally the processor executes the programme from the PROM called the resident programme. When an unanticipated problem occurs onboard, a new programme can be developed on the ground and transmitted to the OBC. The OBC would then be commanded to perform operations corresponding to the newly transmitted programme. Apart from changing the entire programme, minor modifications in the existing programme such as changing gains or addition of a new mode called dummy mode can also be carried out.

A non-maskable interrupt (NMI) is a hardware interrupt that cannot be ignored by standard interrupt masking techniques in the system. In the OBC, if the watchdog timer crosses a specified value, an NMI is generated which points to NMI vector (i.e. start from minor cycle 1).

8.2.3 OBC software design

The design follows either a waterfall model or a component-based software development methodology wherein each module is developed and tested individually. Once all components are developed they are incrementally integrated and tested. The entire OBC software is developed as one major cycle, which is divided into a number of minor cycles. During each minor cycle, the OBC processes separate sub-systems and there should be a positive time margin after completion of each major cycle. The software languages used normally are ADA or extended C.

The typical flow of OBC software is shown in **Fig. 8.2.**

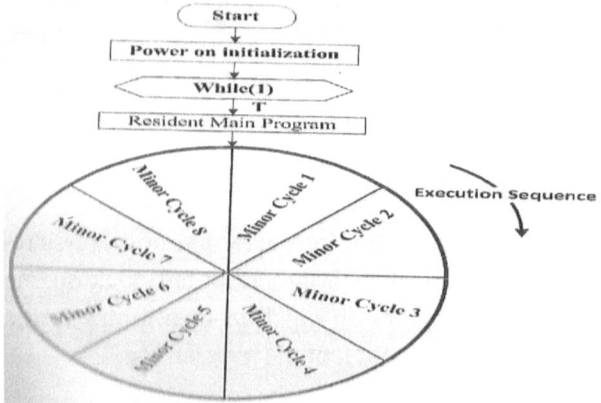

▲ **Fig. 8.2:** OBC S/W cycle diagram

Once the system is turned on, the first function to be executed is the 'power ON' initialisation, which makes the necessary initialisation of the CPU and IO. This is followed by the resident main programme. The execution stays in the resident main programme, which contains all functional requirements. As the execution enters the resident main programme, it follows a cycle. All the models under the resident main programme fall in one of the minor cycles. The execution starts from the minor cycle number one and ends with minor cycle number eight and this process repeats. The entire OBC software has to be developed as one major cycle, which takes 128 milliseconds (typically) and is hence divided into eight minor cycles with one minor cycle taking16 milliseconds (typically) to complete its task.

For important parameters, a majority voting logic (MVL) is incorporated wherein parameters are stored in three locations in the memory. During a reading, parameters from all the three locations are accessed and data matching from at least two locations is taken for further processing.

To take care of single event upset in the memory, especially RAM, an error detection and correction logic (EDAC) is incorporated for the RAM. For every 16 bits of data, six additional parity check bits as per modified hamming code are generated and written into the RAM. While reading the data from the RAM, using the six additional parity check bits, error detection and correction (single bit) are carried out for the 16-bit data and the corrected 16-bit data is used for further application. Two-bit errors can be detected and no correction is done.

Latest developments in miniaturisation have resulted in developing an OBC as a single board system using a 32-bit processor, operating at 180 MHz, utilising 2 MB flash memory and 512 KB SRAM. It supports interfaces such as SPI, I2C, UART and GPIO. It has been developed by ISRO for its ISRO Nanosatellite (INS) and is shown in **Fig. 8.3**.

Above mentioned functions are developed using separate sub-system hardware packages in big satellites since the requirements of telecommands, telemetry channels, temperature sensors and heaters are very large in number. In a Nanosatellite, these requirements are generally less in number and these functions are implemented by an OBC with a micro-controller based Firmware with associated digital electronics and interfaces. A brief description of these functional sub-systems, their design considerations and implementation aspects are given in the following paragraphs which are common to all satellites.

▲ **Fig. 8.3:** OBC of INS (Courtesy: ISRO)

8.3 TELEMETRY

Telemetry (TM) system provides data on the health parameters of a satellite through an RF downlink from a Nanosatellite to the ground station. These parameters include voltages, currents, temperatures, ON/OFF relay status and satellite attitude measured by sensors, which are used to evaluate the satellite's performance. Sometimes, the satellite payload data is also transmitted on the telemetry link multiplexed with health data or separately. In a Nanosatellite, telemetry data transmission begins after the satellite is ejected from the launch vehicle and it is powered ON. Typical TM downlink block diagram is shown in **Fig. 8.4**

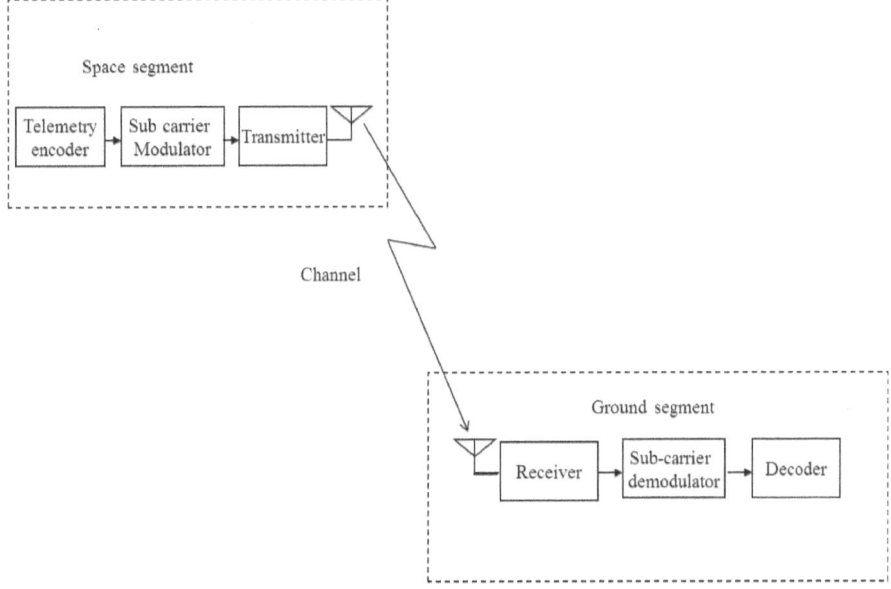

▲ **Fig. 8.4:** Downlink block diagram

Nanosatellite health parameters are sampled at various intervals as per the mission requirements – typically from 128 milliseconds to two seconds. In general, parameters (such as temperatures or battery voltage) that do not change rapidly are sampled at low intervals and vice-versa. These sampled parameters are converted into eight-bit words and arranged into a pre-defined format. The TM data format can be of two types, (i) Normal format: where all parameters with different sampling rates are covered within one master frame of telemetry; and (ii) Dwell format: where a few selected parameters are only filled in the telemetry frame ensuring the highest possible sampling rate. The parameters that are to be dwelled are command selectable so that different parameters can be dwelled at different phases of the mission.

A TM frame has a frame sync code (FSC) as a frame header. The frame sync identifies the start of a frame and in some cases, it is used to validate the quality of data received in the frame. The length of the frame sync code is given by

$$\text{Length of FSC (L)} = (2 \text{ to } 10) \times \text{Log}_2 N$$

where L is the length of the FSC and N is the number of bits in a TM frame. For example, if a TM frame length is 1,024 bits, the FSC length will be a minimum of 20 bits and a maximum of 100 bits. A truncated pseudo-noise (PN) sequence is used as FSC because of the random properties of the PN sequence.

8.3.1 PN Sequence Properties

- PN sequence length is $2^N - 1$ (where N is number of stages in the shift register)
- Balance property: The difference between the number of 1s and 0s in the code will be 1
- Run property: All combinations of 1s & 0s are possible. Half the runs are of length 1 (010, 101), ¼ the runs are of length 2 (0110, 1001)
- Shift & add property: A resultant of a shift and add of PN sequence is also a PN sequence
- Autocorrelation property: PN sequence can be correlated only by itself

After the FSC, the TM frame consists of one or two words to represent a Nanosatellite's identification. If all parameters with their sampling rates cannot be covered in a single frame, the TM format is arranged in multiple sub-frames typically 16 or 32. Hence one or two words are allocated in TM frames to identify the sub-frame ID numbers and type of TM data format (normal or dwell).

A typical TM frame format and how different sampling rates are adjusted in a frame are shown in **Figs. 8.5 and 8.6.**

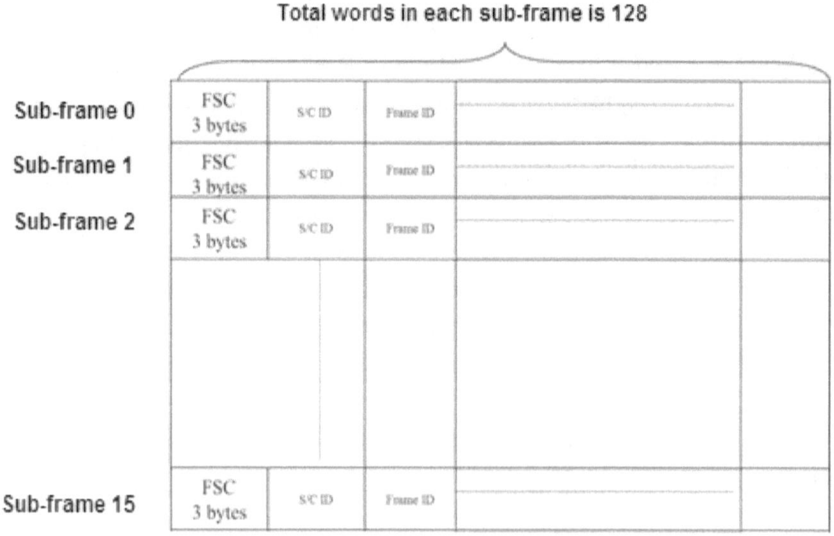

▲ **Fig. 8.5:** Format of normal telemetry

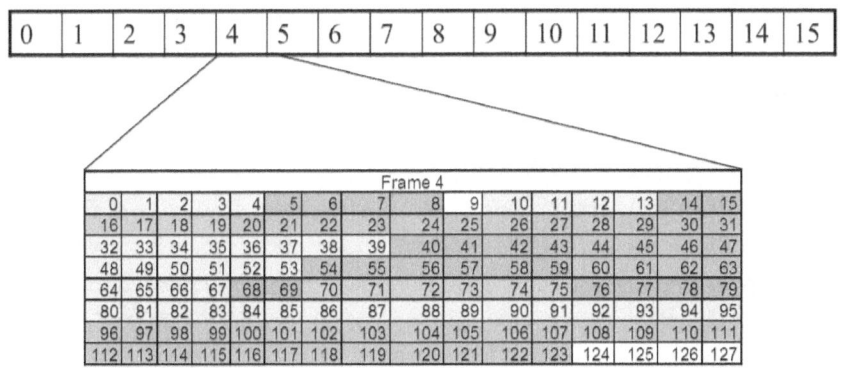

colour code	sampling	number	No. of words
	1s sampling	43	688
	2s sampling	32	256
	4s sampling	24	96
	8s sampling	26	52
	16s sampling	874	874
	Spare	82	82

Word number	Data value
Word 0	FSC word#1
Word 1	FSC word#2
Word 2	FSC word#3
Word 3	Satellite ID
Word 4	Frame ID

▲ **Fig. 8.6:** Typical telemetry format

8.3.2   TM data acquisition

The health parameters of a satellite are of various types and their acquisition techniques used in telemetry system are as follows:

a) Analogue channels

They are in the range of - 5 volts to + 5 volts normally. The analogue channel is converted to digital eight bits with a resolution of 40 mV. In some special cases, even 10-bit resolution is also provided for some critical parameters. In order to reduce the no of ADCs (analogue to digital converters), all the analogue parameters are multiplexed to get a single channel at a time so that a single ADC can be used to convert all analogue channels, one at a time.

b) Digital channels

They are zero or + 5-volt monitoring. In order to reduce the interface lines, they can also be serialised using digital multiplexers and serial digital interface of digital words of eight bits, 16 bits or 64 bits.

c) RF bit micro-switches monitoring

These are switches to select main and redundant RF system outputs and event monitoring using micro switches. They provide two contacts which are open in one condition and closed in other conditions. Hence TM provides signal conditioning for these to convert open/close status to 5 volt/0 volt and make them digital bit monitoring.

d) Temperature sensor monitoring

Different types of temperature sensors such as thermistor, platinum resistance thermistor (PRT), thermocouples or fine temperature sensors are used in a satellite depending on the temperature range and sensitivity needed in the sub-systems. The signal conditioning to convert the temperature data into analogue voltage monitoring is carried out by telemetry.

8.3.3   Onboard time (OBT)

In order to have a common time reference for various events that happen in a Nanosatellite, the TM incorporates onboard time which is a crystal clock derived timer. This time is used when referring to all onboard events and used as a reference time for commanding, controlling events and payload events. The OBT will have a drift correction and offset adjustment to make it more accurate. TM formats will sample this OBT in every frame so that this time is available for ground processing.

8.3.4   TM format and sub-carrier modulation

Various PCM data formats are shown in **Fig. 8.7**. The TM data after formatting is in the form of PCM NRZ-L. As this data format does not ensure transitions when continuous zeros or ones are present in the data, it is standard to convert NRZ-L to NRZ-S before RF modulation. This NRZ-S makes sure that there are transitions in the data when continuous zeros are present in the data. Normally as TM data rate is very low, of the order of one Kbps or four Kbps, it is not possible to modulate this low data on the main carrier with a frequency in the VHF Band/S-Band.

Code Name	Binary Code	Code Definition
NRZ-L	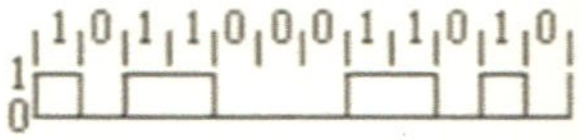	Non-Return-to-Zero Level "One" is represented by one level "Zero" is represented by another level lower the one but not zero.
NRZ-M	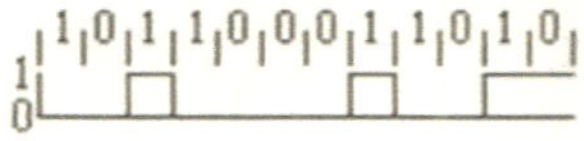	Non-Return-to-Zero Mark "One" is represented by a change in level "Zero" is represented by no change in level
NRZ-S	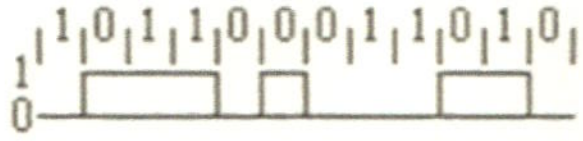	Non-Return-to-Zero Space "One" is represented by no change in level "Zero" is represented by change in level.
Bi-Phase-L	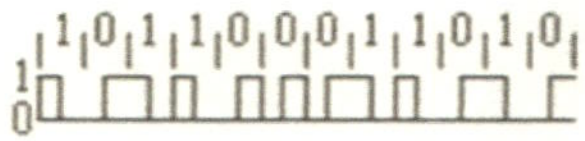	Bi-Phase Level (Split Phase) Level change occurs at the beginning of every bit period "One" is represented by a "One" level with transition to the "Zero" level "Zero" is represented by a "Zero" level with transition to the "One" level
Bi-Phase-M	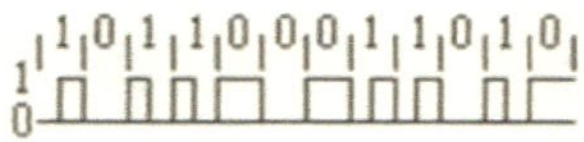	Bi-Phase Mark Level change occurs at the beginning of every bit period "One" is represented by a midbit level change "Zero" is represented by no midbit level change.
Bi-Phase-S		Bi-Phase Space Level change occurs at the beginning of every bit period "One" is represented by no midbit level change "Zero" is represented by a midbit level change.

▲ Fig. 8.7: PCM data formats

Hence sub-carrier modulation is used to translate this TM data to 32 kHz or 128 kHz using PSK (phase shift keying) technique. One more advantage of using a sub-carrier modulation is that the final RF modulation will be analogue (mostly phase modulation) so that at the ground station, the carrier is always present during visibility to ease data reception. With sub-carrier modulation available, a second formatted data such as dwell data or playback data can also be simultaneously transmitted using different sub-carriers modulating the same main carrier. This is also useful in carrying out simultaneous TM and tone ranging at the ground station. In some Nanosatellites, TM data is modulated on a beacon to avoid a separate transmitter.

8.3.5 Onboard data storage

LEO Nanosatellites are not continuously visible to the ground station. Hence an onboard data storage system is provided to record the TM data during the non-visibility period. Depending on the number of non-visible orbits, the TM data can be stored either in a continuous or in sub-sampled mode. This stored TM data is played back (PB) at faster rates when the satellite is visible to the ground station.

There are two ways of sending playback data:
1. Send playback (PB) data on a second sub-carrier so that it can be available simultaneously with real-time normal TM data.
2. Playback (PB) data can be multiplexed with normal data so that on a single data stream both data are available. On the ground, they can be de-multiplexed and separated into two different data streams.

This scheme is shown in **Fig. 8.8**
Real-time data

RT1	RT2	RT3	RT4	RT5	RT6	RT7	RT8	RT9	-	-	-

Storage & real-time data

ST1	ST2	ST3	ST4	ST5	ST6	ST7	ST8	ST9	RT1	ST10	ST11

▲ **Fig. 8.8:** Real-time & PB TM format

Typical TM system specifications are given in Table. 8.1

▼ **Table. 8.1:** Typical TM system specifications

Parameter	Specification
Frame length	128 words
Word length	8 bits/word
Normal TM sampling rate	128 m sec/1 sec/2 sec/4 sec/8 sec/16 sec
Dwell TM sampling rate	8 m sec
Bit rate Normal TM	1 Kbps/4 Kbps with 32 kHz sub-carrier
Dwell TM	1 Kbps/4 Kbps with 128 kHz sub-carrier
Playback TM	16 Kbps with 128 kHz sub-carrier
Frame sync code	24 bits (truncated PN sequence)
Frame ID word	6 bits
No. of sub-frames / frame	16/32
Output	4V peak-to-peak (NRZ-S) PSK
Onboard storage	3 M bytes (typical)
Onboard timer	32 bit counter with 1 msec/8 msec resolution

8.4 TELECOMMAND

The main function of the telecommand system of a Nanosatellite is to operate and control the satellite by a set of commands sent from the ground station to the satellite. This is accomplished by transmitting coded instructions from the ground station over an RF carrier referred to as the uplink to the Nanosatellite's receiving equipment. The Nanosatellite is controlled via commands to switch payload ON/OFF, change the orientation of the satellite or change gain settings. A typical block diagram of command uplink is shown in **Fig. 8.9**.

Commands can be classified as single commands and block commands. Single commands are employed when controlling specific functions. Single commands are a command set that is equal to one set of binary digits, which can cause only one function to be performed on the satellite. A block command represents a number of single commands, which will be transmitted in a specific order.

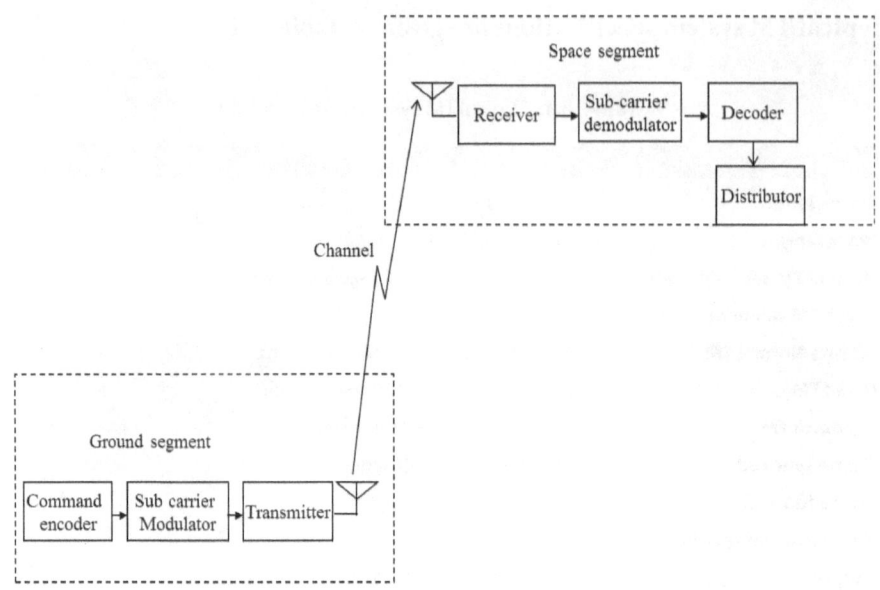

▲ **Fig. 8.9:** Uplink block diagram

Commands can be further classified as real-time (RT) commands or time tag (TT) commands. The primary difference between these commands is the time of execution. A real-time command is executed immediately after receiving the command whereas a TT command is executed after a preset time delay after receiving the command. However, both RT and TT commands are uplinked to the satellite when it is in the radio visibility of a ground station. Some commands are pre-programmed with onboard logics to operate in the autonomous mode like a heater will switch ON/OFF based on the programmed upper and lower temperature limits.

Broadly there are two types of commands:
1. ON/OFF commands
2. Data commands

In the ON/OFF command, any time the command is received, it will perform the same specified function. For a data command, the TC provides the data pattern as needed by the sub-system. This data is user (sub-system) selectable and it is variable.

8.4.1 Command code selection

A telecommand link requires a very low probability of wrong command execution and right command rejection. As command information is transmitted in bursts, block codes are the most suitable. BCH (Bose Choudary and Hockengum) codes have the following properties which make them the best suited for a telecommand link:

1. Binary linear cyclic code
2. Well understood algebraic structure
3. Well-developed algorithms for implementation
4. Good hamming distance for given information and block length

Hamming distance is defined as the number of places in which two codes differ. Hamming weight is the number of ones in a code. For linear codes, the minimum hamming distance is the minimum weight of the code. The hamming distance determines the number of errors that can be detected and corrected in a code.

Depending on the command information length, a BCH code of suitable length will be selected. However as BCH codes are cyclic codes (meaning any cyclic shift of the code results in a valid code and cannot be detected by the decoding logic), the original code is shortened by a few bits to make it non-cyclic. The shortened BCH code is mostly selected for a telecommand system. One of the important parameters of the code selection is hamming distance. Depending on the environment (noise level present in the channel), the sensitivity of the RF receiver and use of error detection and correction, suitable hamming distance is chosen. A single error correcting BCH code is also known as a hamming code and has a hamming distance of 3.

An example of BCH code used in telecommand is (63, 39, 24, 9) shortened to (56,32, 24,9) where 63 bits is the length of code, 39 bits are information bits, 24 bits are the parity check bits and 9 is the Hamming distance. After shortening by seven bits, 56 bits is the length of code, 32 bits for information, 24 bits are parity check bits and 9 is the hamming distance. As it can be seen shortening doesn't reduce the hamming distance, it only reduces the information capacity of the code.

8.4.2 Command format

After finalising the command requirements, the command format has to be worked out. The first bit of the command format will be the start bit which will be normally 1. Then a few bits will be allotted for the satellite's address (ID) which specifies the command as meant for a specific satellite. The command is executed only if this address matches, otherwise it will be rejected. The next field will be types of commands such as ON/OFF, data, real-time or time tag. The last field will be the command information and command address. Having arrived at the command format, the entire command information is coded using a BCH code and the parity check bits are appended to the command bits. This is called a command frame. Normally, to increase the command execution probability, the command frame is repeated two times or four times with a gap between frames. A typical telecommand frame is shown in **Figure 8.10**.

1	SC ID	LinkID	Mode	Command Information	
B1 B2		B6 B7	B8 B9	B16 B17	B32

▲ **Fig. 8.10:** Typical telecommand format

B1	:	Start bit (Always 1)
B2 to B6	:	(Spacecraft ID)
B7 to B8	:	Link ID (01 – Link#1, 10 – Link#2)
B9 to B16	:	Mode
B17 to B32	:	Command code/data

A time tag command transmission requires two commands - first, it will indicate in the command format that the command is time-tagged and hence it has to be executed later. Second is the time when the previously uplinked command is to be executed. The telecommand decoder, on detecting the time tag command, stores the command information and time tag. When the predetermined time arrives, it will retrieve the command from the storage and execute it. Some of the additional features incorporated in the time tag logic are:

1. Edit the command or delay or both. Automatic queuing of time tag commands. Rearranging them in the order of time of execution.
2. Block execution of time tag commands where one can specify the start address and stop address of the time tag stack and time tag commands in that window will get executed on maturity.
3. Normally time tag commands are referred to the OBT in the time delay. In some cases, the time tag can be referred to the ground time i.e., the time at which the time tag command is uplinked from the ground.

These functions are normally configured with the telecommand processor (TCP) in the onboard computer.

8.4.3   TC sub-carrier modulation

As telecommand data rate is very low (100Bps, 1Kbps or 4Kbps), it is not directly modulated on the main carrier. Also, in order to facilitate simultaneous commanding and ranging, the command data is translated into the frequency using the sub-carrier modulation. The sub-carrier frequency is selected such that it does not interfere with the ranging tones. Usually, two types of subcarrier modulation are used in a TC:

1.   Frequency shift keying (FSK)
2.   Phase shift keying (PSK)

In FSK, two sub-carrier frequencies are selected. One frequency (f_1) for 0s and the second frequency (f_2) for 1s. In PSK, a single sub-carrier frequency with a 0-degree phase for 0s and 180-degree phase for 1s is used. The advantage of an FSK is that it can be demodulated by non-coherent techniques which are simple and easy to implement. As sub-carrier demodulators in a TC have to be incorporated onboard the satellite, simple, low power consuming and reliable techniques are used. However, in terms of performance in the presence of noise, PSK is better than FSK. But PSK demodulation has to be carried out by coherent demodulation only using PLL techniques, which are more complex compared to FSK demodulation. The noise performance of different digital communication techniques is shown in **Fig. 8.11**.

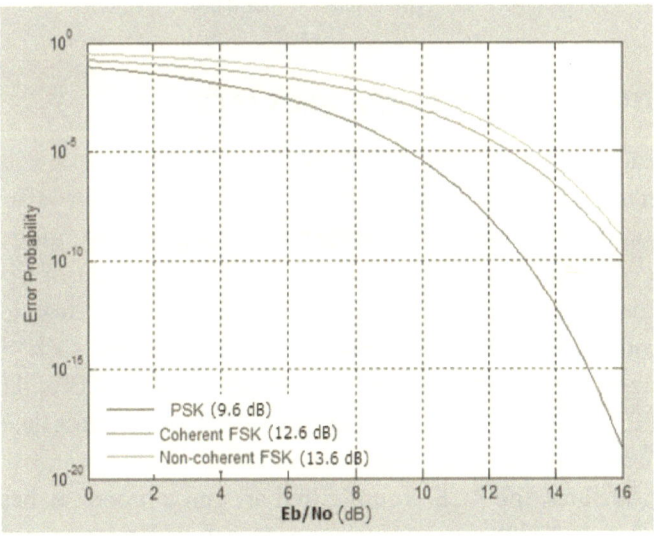

▲ **Figure 8.11:** BER vs Eb/No. of modulation schemes
(Courtesy: Principles of communications by Taub and Shilling)

For a telecommand link, noise performance is always compared for a bit error probability of 1×10^{-5} (i.e., one-bit error in 10^5 bits). Thus, it can be seen from the figure that PSK modulation gives Eb/No (signal energy per bit/ noise energy per cycle) of 9.6 dB. The similar figure for a non-coherent FSK is 13.6 dB and for coherent FSK, it is 12.6 dB.

A typical block diagram of a telecommand system is shown in **Fig. 8. 12.**

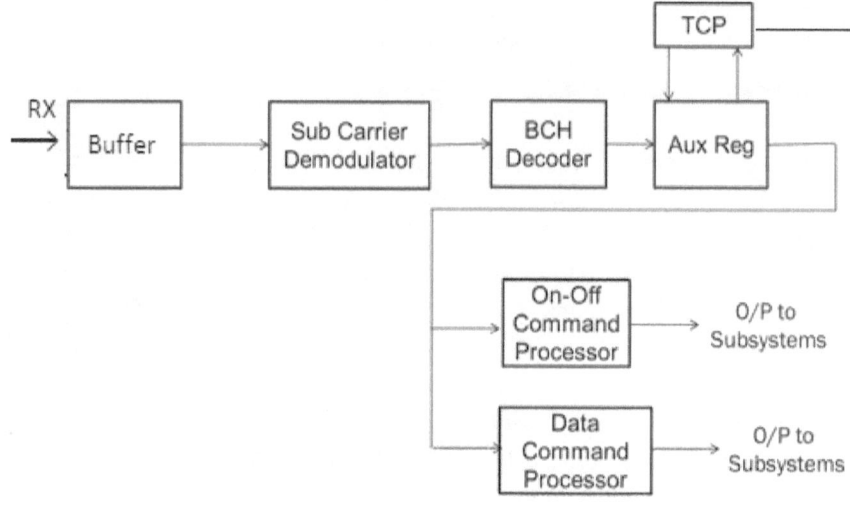

▲ **Fig. 8.12:** Typical telecommand block diagram

8.5 ATTITUDE CONTROL ELECTRONICS (ACE)

Another major function of an OBC is Nanosatellite attitude control. Attitude control electronics (ACE) receives attitude information from various sensors (sun sensor, star sensor, magnetometer), processes the information in accordance with control algorithms and the chosen mode of operation to generate control signals for actuators (magnetic torque, reaction wheel). Attitude control systems are highly mission dependent such as payload pointing, stability requirements, mission orbital characteristics and the control system stability as well as response time. More details on attitude determination and control are covered in the chapter on ADCS.

In big satellites, the ACE is configured around a processor-based system using a 16-bit or 32-bit microprocessor with full redundancy and sufficient PROM and RAM storage capacity. The CPU has its programmes stored in the PROM and the RAM is used for intermediate data storage and remote programming. The clock frequency at which the CPU works is decided from the software and maximum read-write loop delay in the circuit and various control algorithmic requirements to be met with. MIL-STD-1553B or other standard interface is adopted for data transfer between various sub-systems and the ACE to have a standard protocol and reduce hardware requirements. In Nanosatellites, due to size and power constraints, all functions of the ACE are implemented in the OBC along with FPGAs and other peripheral devices. It will have interfaces with attitude sensors to get attitude errors (roll, pitch, yaw) and with inertial measurement unit (IMU) to get body rates. Based on the attitude determination algorithms implemented, control signals are given to magnetic torquers and reaction wheels to correct attitude errors within limits. Serial to peripheral interface (SPI) and I2C interface are used with these systems to transfer data. A typical interface of an OBC with a sun sensor is shown in **Fig. 8.13**.

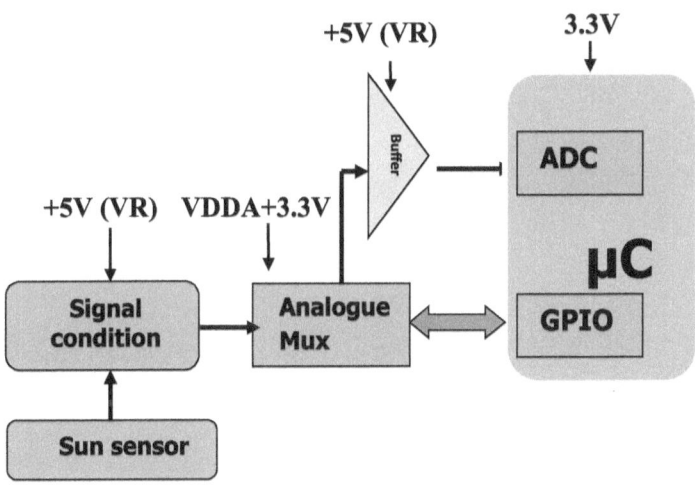

▲ **Fig. 8.13:** OBC interface with Sun sensor (Courtesy: ISRO)

To improve the performance of ACE, fault tolerance features such as fault detection and isolation (FDI), safe mode logic, remote programming, error detection and correction logic (EDAC) for RAM and majority voting logic (MVL) for a few critical parameters are provided.

ACE implements the following modes of control operations – De-tumbling mode, three axes acquisition mode, normal mode and safe mode. These modes of operation are explained in Chapter-10.

REFERENCES

1. ESA TM & TC Standards
2. Error-correcting codes by Shu Lin
3. Error-correcting codes by Peterson
4. *'Principles of Communications'* by Taube and Shilling

* * *

K. Parameswaran is a satellite digital systems expert, who has more than three decades of experience in design and development of digital electronics and on-board computers for satellite telemetry, telecommand, thermal control management and attitude control systems. He made significant contribution in the design and development of an integrated On-board computer for Indian satellites.

COMMUNICATION SYSTEM AND GROUND SEGMENT

Manohar Sonnada & Raghavendra Hathwar

Generally, small satellites, including Nanosatellites, are placed in low earth orbit due to which satellite radio visibility to any ground station is limited to around 15 minutes. The communication system, also referred to as an RF system, provides the link between satellites - the space segment and earth station, which is a part ground segment.

The space segment configuration depends on factors like the orbit, application, power availability, frequency and bandwidth. The ground segment configuration depends on earth station capabilities, link margin requirements, visibility duration of satellite covering duplex communication between space and the ground.

The communication systems can be broadly classified into telemetry, tracking and command (TTC) as well as payload data transmission and reception. Due to low data rates, TTC communication systems generally use S-band or C-band, whereas payload data transmission needs high data rates for which higher frequency bands are used.

In case of Nanosatellites, the International Telecommunications Union (ITU), which is the authority to regulate satellite frequencies, has allotted VHF/UHF bands for both TTC and payload data with innovative digital coding and modulation techniques.

The reader can get a fine understanding of this technology in this chapter.

Communication system plays a pivotal role in satellite operations. The overall communication system is distributed between i) the space segment comprising telemetry, tracking and command (TTC) system and mission-specific payload data transmitters, which reside in the satellite; and ii) the ground segment comprising a network of ground stations, satellite control centre and mission control centre. The communication system provides a link between the satellite and the ground station.

Most of the Small/Nanosatellites are placed in low earth orbit (altitudes of 400 to 1000 kms) or polar orbit with an orbital period of 85 - 100 minutes. As the satellite keeps orbiting, its radio visibility duration over a ground station is limited to 12 to 20 minutes. Hence communication systems have to be configured to transmit data to the ground and receive signals from the ground to complete all functions such as command reception, telemetry and payload data download during this visibility period. Sometimes, many ground stations are needed to be networked to increase ground contact/radio visibility time with suitable mission planning.

Communication systems in Nanosatellites are going through an evolution. Early satellites carried simple beacon transmitters which could downlink a limited amount of telemetry data. In recent times, Nanosatellites carrying application-specific payloads such as space science research, weather monitoring, science experiments or earth observation missions that demand higher data throughput. Traditionally the communication between the satellite and the ground station employs the radio frequency (RF) spectrum with maximum data rates of around 10 Mbps depending on frequency and modulation, although experimental laser/optical communication systems have shown promises of data rates up to 200 Mbps.

Later in the chapter, the necessity for laser/optical communication, its advantages, onboard system configuration and optical ground station configuration have also been briefly described.

Typically, the RF communication system comprises a receiver, a transmitter and an antenna system in both the space segment and the ground segment. A brief description of the communication system elements is covered in the subsequent paragraphs.

9.1 SPACE SEGMENT

The space segment communication system configuration depends on many factors such as the orbit, application, power availability, frequency and bandwidth, ground station capability and network, link margin requirements as well as radio visibility duration.

9.1.1 RF Communication

a. **Frequency selection**
Satellite communication systems can be broadly be classified into (i) Telemetry, tracking and command (TTC); and (ii) Payload data transmission. Due to low

data rates, TTC communications generally use S-band or C-band whereas payload data transmission (particularly earth observation payloads) need high data rates where high-frequency bands such as X, Ku or Ka bands are employed. Most of the Nanosatellites are configured in the amateur VHF/UHF frequency bands due to no frequency co-ordination requirements and readily available COTS (commercially-off-the-shelf) version flexible (settable frequency, power) communication modules. However, due to limited bandwidth availability in amateur spectrum in VHF/UHF bands, the designers are forced to adopt/develop systems in higher amateur/commercial S/X frequency bands for higher data rate payloads. The usage of commercial frequency bands needs to undergo frequency allocation/coordination process with ITU and other satellite operators, which is a complex and time-consuming process. Also, adopting higher frequency bands for Nanosatellites is associated with technical complexities such as high power requirement, higher thermal heat dissipation, complex antenna design in addition to higher cost. Available amateur frequency bands for Nanosatellites are given in **Annexure- A4**.

b. Transmitter & receiver (transceiver)

Telecommand receiver and telemetry transmitter in most Nanosatellites operate in VHF/UHF bands due to legacy and availability of COTS version modules on a single PCB. This module receives telecommand signals from the ground and sends the satellite health data as telemetry to the ground station. The module can support different kinds of modulation/demodulation schemes for the data rate from 0.5 Kbps to 120 Kbps. The transceiver module supports standard HDLC/AX.25 protocol/raw pattern of data formats for both uplink and downlink communication.

The telecommand receiver has a sensitivity of around -115 dBm and operates in the dynamic range of -115 dBm to -50 dBm. Local oscillator signals for receiver and transmitter are generated from the on-chip PLL and TCXO available in the module. The transmitter output power is generally adjustable typically up to 27 dBm in steps of 0.5 dB.

c. Payload data transmitter

Payload data transmitters are application-specific and generally operate in high-frequency bands L, S, X or Ku/Ka due to high data rates and higher available bandwidths. The RF power output will be in the range of 30 dBm (decibel milliwatts). Both frequency and power output are programmable. The forward output power from the power amplifier is coupled through a directional coupler and monitored. The transmitter receives the payload formatted data from the onboard computer (OBC) for further modulation and transmission. The modulation scheme adopted is usually bi-phase shift keying (BPSK)/Quadrature phase shift keying (QPSK) with data rates of up to 1 Mbps and sometimes even more, depending on the frequency of operation. Other modulation schemes can also be used depending on the requirements within the satellite resource constraints. The data downlink transmission can be continuous or in burst mode depending on the onboard power, ground station gain-to-noise temperature ratio (G/T), visibility duration, bandwidth/data rate and the link margin availability.

d. Modulation schemes

Fast-growing interest in Nanosatellites for complex missions has led to increased demand on downlink data rates. Owing to limited onboard resources –mass, volume, power and minimisation of cost – improved miniaturised communication systems in COTS version with spectrally efficient modulation schemes and better channel coding are used. Many satellites still use less efficient AFSK/FSK link but a few missions are employing modulations such as BPSK, QPSK. VHF/UHF links are used up to data rates of 100 Kbps whereas S-band links are used for data rates of up to 1Mbps and sometimes even higher.

e. Antenna system

A transmitting antenna converts electrical signals into electromagnetic waves and radiates them.

A receiving antenna converts electromagnetic waves from the received beam into electrical signals.

In two-way communication, the same antenna can be used for both transmission and reception, if uplink and downlink frequencies are close by, otherwise, two different antennae are needed for better performance.

Mostly the VHF/UHF antennae will be omnidirectional monopole or dipole built with a commercial measuring tape, whereas at higher frequencies the antennae will generally be patch antennae with either linear or circular polarisation. Even deployable helical antennae are used at UHF sometimes. High gain and directive antennae are generally not used due to pointing constraints. For more demanding payload data needs, high-gain antennae such as deployable reflector and inflatable antennae have been developed for Nanosatellites - but their deployment systems and pointing requirements are significantly more complex. Hence generally, near-omnidirectional antennae are used for telemetry tracking and command (TTC) purposes and directional and wide beam antennae are used for payload data transmission.

Link calculation details, including elements of space communication system, various parameters, link equation and preferable margins for a reliable RF link are given in **Annexure-A5**.

9.1.1.1 Communication system configuration

A. **Telemetry and command system**
Typical block schematic of a TTC system is shown in **Fig. 9.1.** In most of the Nanosatellites with VHF uplink and UHF downlink, separate antennae are used for uplink and downlink. A common antenna can be used when uplink and downlink frequencies are close by, like in S/X-band TTC System. The received VHF carrier is passed through a low noise amplifier (LNA), down-converter and the demodulated baseband signal is sent to the OBC for command decoding and execution. Similarly, the telemetry health data is modulated, up-converted to a UHF frequency, amplified and transmitted to the ground station through the antenna.

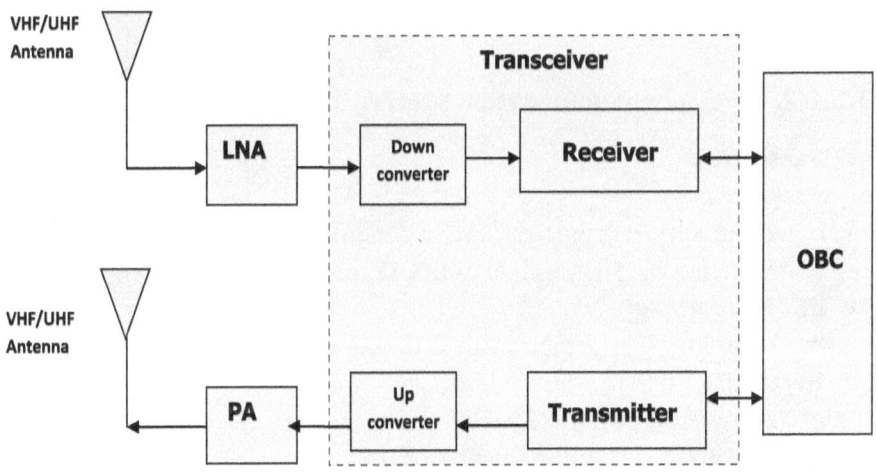

▲ **Fig. 9.1:** TTC Transponder

Nanosatellite tracking requirements are generally met by programme track, use of single/dual-frequency GPS receivers on satellite, monitoring downlink Doppler change and using NORAD TLE as well as orbit propagation.

B. **Payload data transmitter**
Payload data is generally converted to digital data before transmission to a ground station. High frequency (S or X band) carrier is generated using a

temperature-controlled crystal oscillator (TCXO)/frequency synthesiser and payload data is modulated onto the carrier, amplified and transmitted to the ground station through a patch antenna. A typical payload data transmitter block schematic is shown in **Fig. 9.2.**

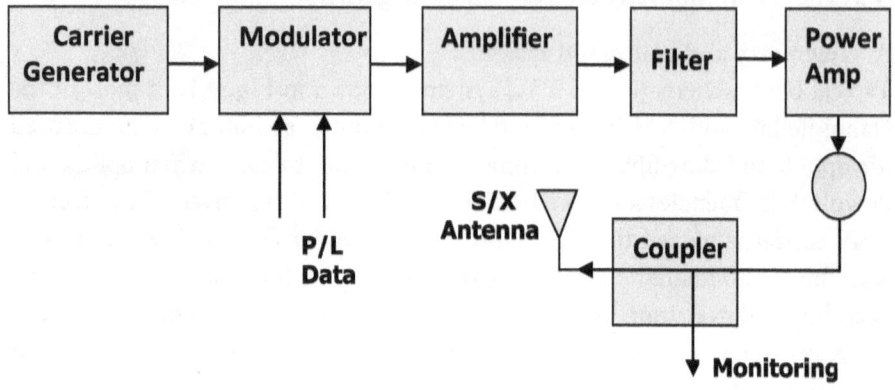

▲ **Fig. 9.2:** Payload data transmitter

9.1.1.2 Typical communication system specifications

A. **Transceiver:**

a. **Package**
   i.   Form Factor: Standard CubeSat Module (PC104 Form)
   ii. Dimensions: 96 mm X 90 mm X 17 mm
   iii. Mass: < 90 g
   iv. Operating Temperature: -15°C to +55°C

b. **Receiver**
   i.   Frequency: 144 to 148 MHz
   ii. Sensitivity: -115 dBm for BER of 10^{-8}
   iii. Noise Figure: < 5 dB
   iv. Spurious Responses: < 65 dB
   v.   Dynamic Range: -115 dBm to -50 dBm
   vi. Frequency Stability: ±3 ppm
   vii. DC Power: < 300 mW/5V

c. **Transmitter**
   i.   Frequency: 420 to 450 MHz
   ii. RF Output Power: +4 dBm to +27 dBm, programmable
   iii. Spurious Responses: < 65 dB

   iv. Frequency Stability: ±3 ppm

   v. DC Power: <2.7 W/5V @ +27 dBm RF Power

d. Processing and Interfaces

   i. Modulation: FSK/AFSK (Uplink/Downlink)

   ii. Data Rate: up to 115 Kbps, selectable

   iii. Communication Interface: UART (2 channels and command) and SPI (1 channel)

   iv. RF Interface: SMA, female, 50 ohms (input/output)

   v. I/O Interface: 40-pin custom interface connector for power and communication interface

   vi. 104 pin PC104 stack through connector

e. Available TMTC Options

   i. Protocol: Raw pattern match using sync word

   ii. Frequency: Custom frequency range/carrier

   iii. Modulations: 4-FSK, GFSK, MSK, GMSK, AFSK, FM, ASK

   iv. Encryption: AES 128 or 256 encryption and scrambling

B. **S-band payload data transmitter**

a. Package

   i. Form factor: Standard CubeSat module (PC104 Form)

   ii. Dimensions: 95.89 mm x 90.17 mm x 16 mm

   iii. Mass: <105 g

   iv. Operating temperature: -15°C to +55°C

b. Input specifications

   i. Input Interface: I2C interface with OBC

   ii. Voltage: 5V and 3.3V

   iii. Current: Max. 950mA at 5V and max. 500mA at 3.3V

c. Output specifications

   i. Output frequency: 2200MHz to 2500 MHz

   ii. RF output power: > 30 dBm, adjustable

   iii. Modulation: BPSK, QPSK

   iv. Data Rate: Up to 1 Mbps, programmable

   v. Power consumption: < 6.5 W

   vi. Interface with OBC: SPI for P/L data, I2C for TMTC

   vii. Framing: CCSDS

   viii. Encoding: Convolution

   ix. Spurious response: > 30 dBc

   x. Side band power level: < -25 dBc

d. Processing and interfaces
  i. RF interface: SMA connector
  ii. I/O interface: Nicomatic connector

A typical commercially available transceiver board for VHF/UHF frequency range is shown in **Fig. 9.3**

▲ **Fig. 9.3:** VHF/UHF Transceiver (Courtesy: www.isispace.nl)

C. Typical VHF/UHF Transceiver specifications
Transmitter
- 145.8 – 146 MHz (amateur- VHF allocation). Other ranges available on request
- Transmit power: 23 dBm
- Modulation options: Binary Phase Shift Keying (BPSK)
- Data rate selectable: 1200, 2400, 4800 and 9600 bps
- Data link layer protocol: AX.25

Receiver
- Frequency range: 435 MHz – 438 MHz
- Modulation: Audio Frequency Shift Keying (AFSK)/FSK
- Data rate: 1200 bps
- Sensitivity: -104 dBm for BER of 1 in 10e-5
- Data link layer protocol: AX.25

Different configurations of VHF/UHF TTC antennae are shown in **Fig. 9.4.**

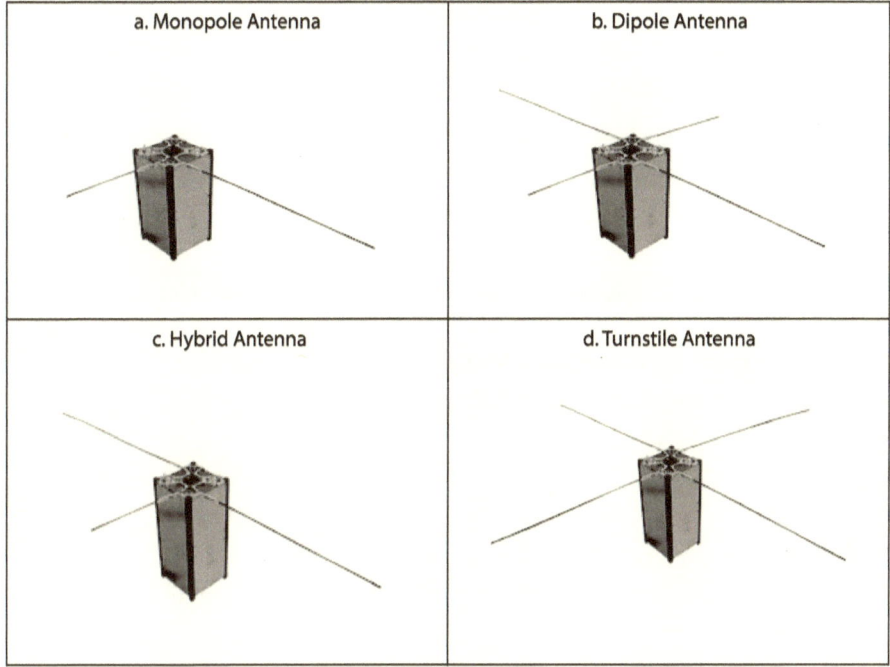

▲ **Fig. 9.4:** VHF/UHF antennae (Courtesy: www.isispace.nl)

These antennae are fabricated with a metallic measuring tape of suitable length. This tape is coiled and held during launch and deployed after the satellite's injection by the launch vehicle.

Typical VHF/UHF antenna specifications and deployment mechanism
- RF impedance (deployed): 50 Ohm
- Max RF power: 2 W
- Insertion loss: < 1.5 dB
- Frequency range: 10-13 MHz bandwidth within 130-500 MHz
- Electrical power nominal: < 20mW, during deployment: 2 W
- Mass (exact mass depends on antenna configuration): < 100 g
- Envelope stowed: (l x w x h): 98 mm x 98 mm x 7 mm
- Supply voltage 3V (5V and 8V available on demand)

A typical patch strip antenna used for S-band transmitter is shown in **Fig. 9.5.**

▲ **Fig. 9.5:** S-band Micro-strip/Patch Antenna (https://n-avionics.com)

Typical patch antenna specifications
- Frequency range: 2238 – 2288 MHz
- Gain: 6 dBi typical
- Vertical beam: ± 40 °
- Horizontal beam: ± 40 °
- F/B Ratio: > 20 dB
- VSWR: < 1.25 for centre frequency, < 1.8 for frequency range
- Impedance: 50 Ω
- Polarisation: RHCP
- Connector: SMA female (straight)
- Outer dimensions: 70 x 70 mm
- Weight: 49 gms
- Isolation RHC/LHC: 12 dB typical at 2263 MHz

A commercial version of the S-band data transmitter configured on a PCB is shown in **Fig. 9.6.**

▲ **Fig. 9.6:** S-band data transmitter (Courtesy: www.isispace.nl)

Typical data transmitter specifications

- Frequency range: 2200-2290 MHz (EESS/SRS/SOS allocations)
- Transmit power: 27 to 33 dBm
- Modulation options: Offset quadrature phase shift keying (OQPSK)
- Pulse shaping: Square root raised cosine, roll-off 0.5, 0.35 (other options on request)
- Channel coding: Concatenated reed Solomon and convolutional coding [C (7, ½) and RS (255, 223)]
- Data rate selectable: 3.4 Mbps (½, ¼ and 1/8)
- Datalink layer protocol: CCSDS

9.2 GROUND SEGMENT – EARTH STATION – RF COMMUNICATION

The ground segment of a satellite mission consists of a single or network of ground stations, a satellite control centre, payload data reception and processing centre, antenna, RF front end hardware such as modems and data processing computers. There will be full-duplex communication between the satellite and the ground station. Due to the orbiting nature of a satellite and limited radio visibility duration, the ground antenna system should have tracking capability to maximise the contact time with the satellite. Thus, the ground station comprises a high gain antenna and its steering/tracking system, Doppler compensating receive chain, transmit chain, station computer for pass scheduling, command operations, acquisition and processing of data, interface with antenna tracking system - using the NORAD two-line elements (TLE) data/onboard GPS data/computer designate mode for tracking and monitoring and control system. All the station systems need to be synchronised with Universal Time (UT, referred to Greenwich Mean Time) and need to have a stable frequency reference source. Hence the ground station will be equipped with a frequency and timing system. Details of NORAD TLE set are given in **Annexure-A8**

Generally, the ground station antenna for Nanosatellites will be a crossed element Yagi antenna in dual/quad array configuration in the VHF/UHF band for TTC functions. If the S-band is used for payload data transmission, a reflector antenna with S-band feed will be used. If the satellite is operating in multi-frequency bands, all antennae will have to be mounted on a single tracking system whenever possible to avoid duplication of tracking systems and also to reduce pointing and synchronising issues. Typical specifications for a Nanosatellite ground station are given in **Table. 9.1.**

▼ **Table. 9.1:** Typical VHF/UHF and S-band ground station specifications and Margins

Parameters	Unit	VHF	UHF	S-band
Antenna	Type	Yagi Uda	Yagi Uda	Parabolic (3m)
Ant gain	dB	16	17	34
Beam width	Deg	22	18	5.0
Rx. NT	K	—	630	500
Rx. G/T	dB/K	—	- 10	7
Polarization	—	Linear	Linear	LHCP and RHCP
U/L Power	W	100	—	—
EIRP	dBW	34	—	—
Data Rate	Kbps	1.2	1.2	19.2/1000
Modulation	—	PCM/FSK	PCM/FSK	PCM/BPSK
ACU Coverage	Deg	AZ and EL AZ 0-360 EL 0-180	AZ and EL AZ 0-360 EL 0-180	AZ and EL AZ 0-360 EL 0-180
Tracking Type	—	TLE	TLE	PGM and TLE
Margins	dB	12-14	5-6	2-3

9.2.1 Ground station configurations

A typical ground station configuration for TTC and payload data reception is shown in **Fig. 9.7** and **Fig. 9.8** respectively. If the uplink and downlink frequencies are widely different, two separate antennae are used as in the case of VHF and UHF. Else, a common antenna feed is used for both uplink and downlink with a diplexer to provide isolation between the two. Generally, ground stations are equipped with discrete units such as low noise amplifiers (LNA), down-converters, modulators/demodulators and data acquisition/processing systems, which increase the station's complexity and cost substantially. In addition, once designed/configured, their usage becomes limited to specific frequency bands and modulation schemes. However, the present trend of implementing a software-defined radio (SDR) based ground station reduces the cost of an operational system significantly without sacrificing the final performance and enables the creation of a more flexible ground station for other applications too. The application and usage of SDR are growing in the wireless communication field to circumvent hardware limitations. This type of ground station can be constantly updated and improved, taking into account technology change and user requirements.

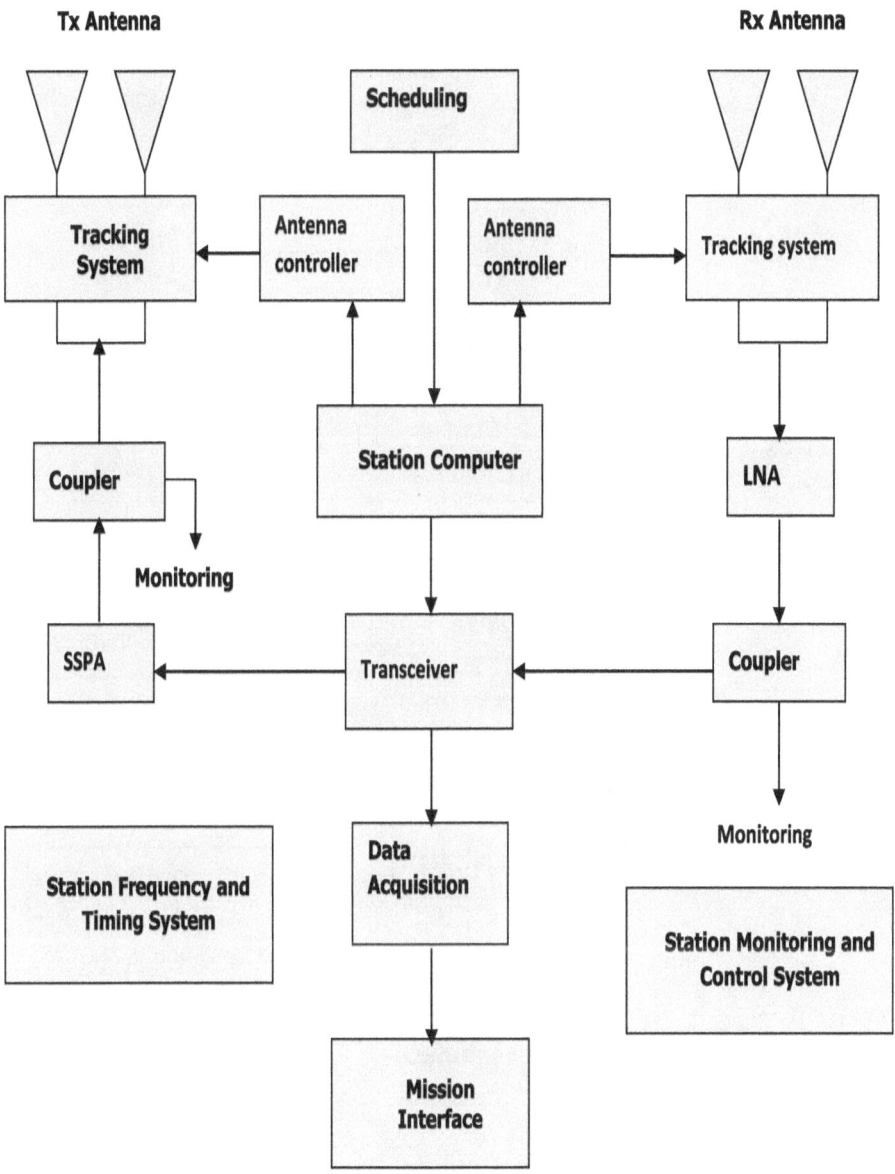

▲ **Fig. 9.7:** TTC ground station configuration

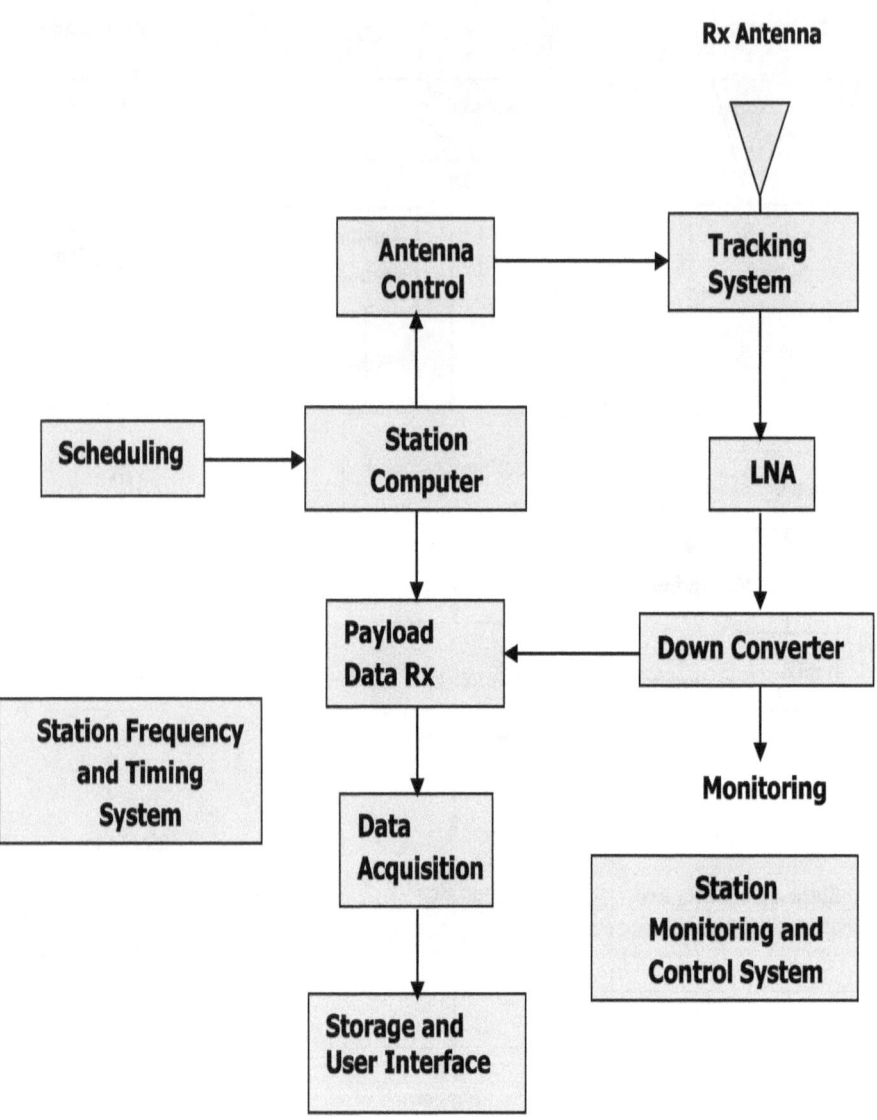

▲ **Fig. 9.8:** Payload ground station configuration

Further information on antenna fundamentals and earth station testing details are explained in **Annexure-A6** and **Annexure-A7.**

9.2.2 Commercially available ground station systems

Typical commercially available VHF/UHF transceiver and S-band payload/TM receiver for ground stations are shown in **Fig. 9.9** and antennae are shown in **Fig. 9.10**

a. VHF/UHF ground transceiver

b. S-band payload/TM receiver

▲ **Fig. 9.9:** Ground station transmitters/receivers (Courtesy: www.isispace.nl)

a. VHF/ UHF Antennas

b. S-band antenna dish   c. S-band and VHF/UHF antenna on a
                                        common pedestal

▲ **Fig. 9.10:** Ground station antennae (Courtesy: www.isispace.nl)

S-band ground station receiver

- Frequency range: 2200 – 2500 MHz (Amateur: 2200 – 2290 MHz or 2400 – 2450 MHz). S-band downlink capability
- Cavity filter to suppress UMTS interference
- Software defined radio receiver, GNU radio compatible
- Flexible data rates: up to 115.2 Kbps
- Noise Figure: <15 dB
- Internal Bias-Tee with switchable 12V DC (1A fused) to power LNA via RF connector
- Graphical user interface software to configure SDR on Debian/GNU LINUX USB 3.0/2.0 connection to a PC
- 85... 264V AC, 50-60 Hz, fused 3.15A input
- 19″ rack-mountable 1U case
- N-type RF connectors
- Packet and control interface using TCP/IP Sockets.
- Support for AX.25 amateur radio protocol. Others upon request

VHF/UHF ground receiver

- VHF frequency range 144 – 146 MHz, other range upon request in 130-170 MHz range
- UHF frequency range 435 – 438 MHz, other range upon request in 400 – 470 MHz range

- 100 Watts (CW) output power in VHF/UHF (possible to reduce the output power to 10 Watts)
- Harmonic suppression: < -50dBc, typical -60dBc
- VHF receiver noise figure: 1.6 dB
- UHF receiver noise figure: 2 dB
- Supply voltage 88 – 264V AC 50 – 60 Hz
- N-type RF connectors
- BNC type for polarization switching output, 0 / +12V DC, 1A max fused
- DC bias for LNAs. 12V via coaxial cable, fused 1A
- Packet and control interface using TCP/IP Sockets.
- Support for AX.25 amateur radio protocol. Others upon request.
- Weight 14kg net
- Software defined radio transceiver
- VHF and UHF downlink capability (In case of commercial bands, additional filters are provided)
- VHF and UHF uplink capability (In case of commercial bands, additional filters are provided)
- Dual-band full duplex
- Supports BPSK downlinks up to 9600 bps on UHF and VHF
- Supports AFSK/FSK uplinks up to 1200 bps on UHF and VHF
- Graphical User Interface Software to configure SDR on Debian/GNU LINUX
- Built-in antenna polarisation switch drive
- Built-in LNA sequencer
- USB 2.0 connection to a PC
- 19″ rack-mountable 3U case

Steerable antenna system
- Azimuth and elevation rotators with speed up to 6°/sec
- Hot-dip galvanised steel mounting mast
- UHF and VHF Yagi antennas
- 2m dish with helix feed, LNA and cavity filters for S-band (2200 – 2290 MHz or 2400 – 2450 MHz)
- Lightning protection system
- 20m of cable between 19″rack and antenna
- S-band Receiver noise figure: 0.9 dB
- VHF antenna: 12.3 dBiC gain, switchable RHCP-LHCP
- UHF antenna: 15.5 dBiC gain, switchable RHCP-LHCP
- S-band antenna: 31.35 dBiC gain
- Power supply: 110V or 230V AC (selectable) 50/60 Hz
  (Courtesy: The above brief specifications are from www.isispace.nl)

9.3 OPTICAL/ LASER COMMUNICATIONS

9.3.1 Introduction

The number of Nanosatellites/CubeSats continue to increase due to their lower cost, lower turnaround time, availability of multiple satellite launch opportunities, the feasibility of faster technology demonstration, rising demand for commercial satellite constellation both for anytime-anywhere communication and earth observation applications.

Also advances in micro and Nanoelectronics as well as high-resolution sensors/cameras have enabled an unprecedented amount of data collection and storage capabilities in satellites. However, sending this amount of data from a power, bandwidth and antenna-gain limited small satellite to the ground is a challenge. Even a good S/X-band CubeSat downlink can handle a data rate of several Mbit/s.

Downloading one TByte data using a single ground station at a rate of 5/10 Mbit/s with an average of four to 10-minute passes per day (very optimistic) would take almost two years.

This ever-growing demand for an increase in data downlink rate and also the congestion in conventionally used radio frequency (RF) spectra (which have been used for decades for satellite communication) have necessitated the need to shift from RF carrier to optical carrier. Optical communications hold the promise of a solution to this problem.

In recent years, free-space optical (FSO) communication has gained significant importance owing to its unique features; large bandwidth, licence-free spectrum, high data rate, easy expandability, easy and quick deployability, less power, smaller size and low mass advantages. The advances in space optics and high power LED technology are also leading towards the development of FSO communication. Optical technology is now being proposed as a substitute for RF equipment even for inter-satellite links (ISL)

The key components of a Nanosatellite/CubeSat optical/lasercom system are the optical-power generation and the pointing capability.

9.3.2 Advantages/disadvantages of optical communication

Free space optical communication has several advantages over traditional radio frequency equipment including:
- Higher data rates, so more information may be transmitted in less time and using lower power
- Better signal/noise ratio (weather-dependent) due to higher directivity

- Lack of interference, better security
- Smaller antennae
- Lower overall power requirements
- Increased spectrum availability
- Narrow beams are difficult to intercept and jam
- No International Telecommunication Union (ITU) coordination needed

On the other hand, there are some drawbacks in optical satellite communications that can cause issues such as:

- Higher pointing accuracy is needed for satellites
- Potential weather-based disruptions
- Increased mission complexity and risk
- The Sun is a noise source for optical detectors

9.3.3 Configuration:

Free-space laser communications use collimated laser beams to transmit information at high data rates in the multigigabit regime, preventing interference problems and exhaustion of RF bandwidths.

Laser diodes (LDs) or light-emitting diodes (LEDs) are used as an optical source in transmitters and photodetectors (PD) are used as receivers.

The first attempt to demonstrate laser communication on a CubeSat, deployed in October 2012 by the robotic arm of the International Space Station (ISS), was onboard FITSAT-1, a 1U system developed at the Fukuoka Institute of Technology in Japan. The satellite carried two arrays of high-power LEDs along with an experimental RF transceiver.

FITSAT-1 used a neodymium magnet as a passive attitude control system with a panel containing 50 green 3W LEDs, achieving 200W pulses and modulated with a 1kHz Morse code signal.

Later many experimental missions have been launched and laser/FSO communication has been experimented/demonstrated.

Laser communication for CubeSats has recently been demonstrated in space but it is quickly maturing. Aerospace Corporation, in cooperation with NASA Ames, launched three CubeSats in its AeroCube Optical Communication and Sensor Demonstration. In March 2018, a systems checkout was completed and the mission entered the operational phase. AeroCube's optical communication technology successfully transmitted 100 MBPS data in August 2018 with AC7B and AC7C CubeSats.

Optical communication system configuration is shown in **Fig. 9.11**

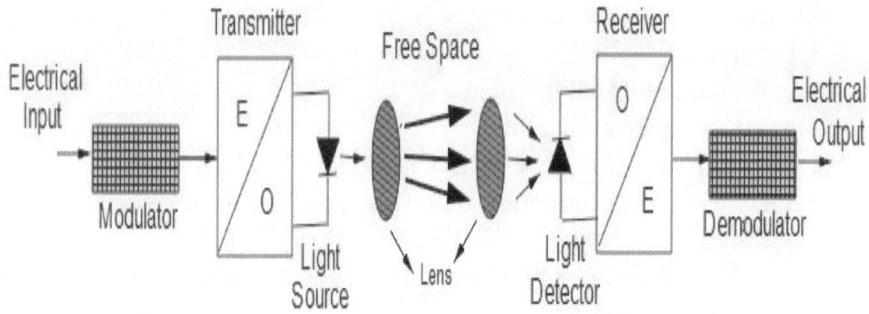

▲ **Fig. 9.11:** Typical FSO communication system schematic (Courtesy: IntechOpen)

Few groups are even working on asymmetric laser communication where the uplink from the ground is simply reflected by a modulating retro-reflector (MRR) on the satellite to overcome the power limitation on satellites and avoid onboard transmitter issues.

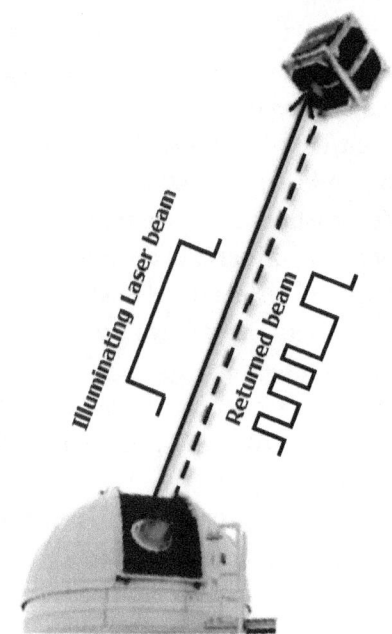

▲ **Fig. 9.12:** Asymmetric Laser Communication scheme
(Courtesy: Communications/NASA)

Free-space optical communication has been identified as an enabling technology for many new space mission concepts but it needs further development to reach its full potential.

9.3.4 Some of the COTS FSO communication satellite sub-systems

a. A lasercom module CubeSat enables a bidirectional space-to-ground communication link between the CubeSat and an optical ground station with downlink speeds of up to 1 Gbps and uplink data rate of 200 Kbps.

▲ **Fig. 9.13:** Lasercom transceiver with built-in attitude determination and control error compensation system (Courtesy: Hyperion Technologies)

It also has an onboard data management system featuring a large data storage buffer and built-in data coding and synchronisation. This makes the CubeCat module a plug-and-play integrated communications sub-system.

b. This laser downlink plus star tracker features a lifetime of 13 years in LEO or nine years in GEO and a downlink data rate of 1 Gbits/s in 1,000-km range, 250 Mbits/s at 2,000-km range and an internal fine-pointing to the ground station based on the built-in star tracker and a passive ground receiver with a 0.55m diameter Newtonian telescope with silicon APD.

The star tracker performance is 5-arcsecond cross-boresight (RMS) and 55 arcseconds around boresight (RMS), and the product has a mass of 335g and a small physical volume of 79x68x68mm

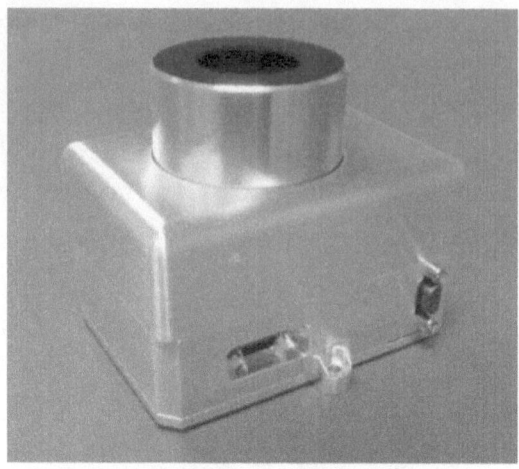

▲ **Fig. 9.14:** The laser downlink, with a built-in star tracker
(Courtesy: Sinclair Interplanetary)

9.3.5 Optical ground station

Optical ground terminals for optical communications are mostly designed for bidirectional links. They comprise both a transmitter and a receiver that generally share the optical antenna. Another peculiarity is the necessity of beam steering with submicroradian angular resolution. These requirements lead to a transceiver block diagram as shown in **Fig. 9.15**

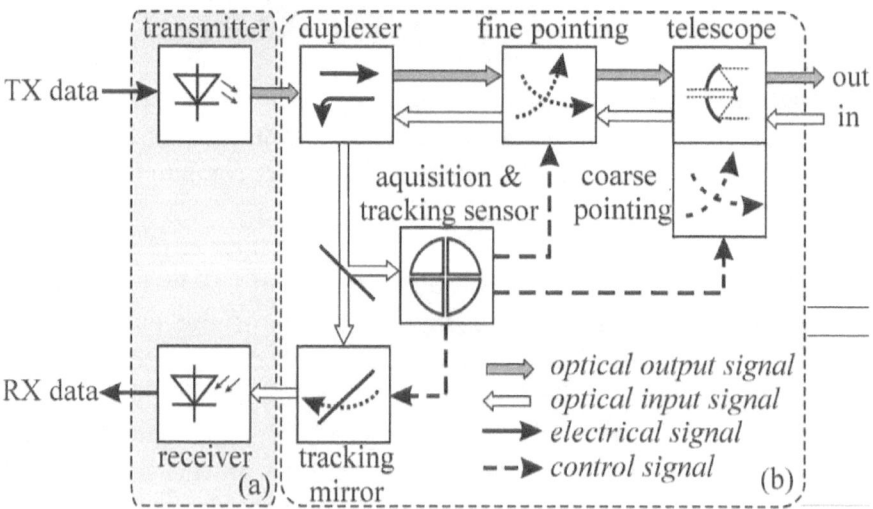

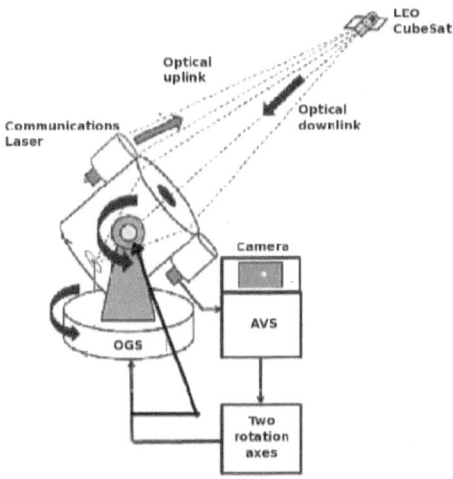

▲ **Fig. 9.15:** Typical optical ground station configuration (Courtesy: Researchgate.net)

The optical ground terminal used for Aerocube satellites: AC7B & AC7C consist of 40-cm diameter, 3-m focal length Ritchey–Chrétien telescope mounted on an AZ/EL gimbal that incorporates two rotary stages. The gimbal is controlled by flight software, which interfaces with servo control software.

The Si-APD detector converts the optical signal into an electrical waveform that is decoded and analysed by an in-house built modem and software package.

Trajectory files for the ground telescope are generated from the vehicle GPS data that is telemetered to one of several RF ground stations prior to the optical engagements. An optical ground station used for AeroCubesat is shown in **Fig. 9.16.**

▲ **Fig. 9.16:** OGS used during AC7B & AC7C Cubesats experiments

Presently, professional telescopes at observatory-class facilities are used for optical communication with satellites.

Making optical ground stations more affordable & transportable is a key enabler for expanding optical/lasercom to small/Nanosatellites and new applications as well as establishing networks to mitigate the effects of weather.

REFERENCES

1. *'Communications'* – Dennis Roddy, Publisher: Tata McGraw Hill Education Pvt Ltd, New Delhi
2. Literature sourced from the internet
3. *'Antenna and Radio Wave Propagation'* by R.E. Collin – Publisher: McGraw-Hill International Editions
4. *'Microwave Engineering'* by Annapurna Das and Sisir K. Das –Publisher: Tata McGraw Hill Education Pvt Ltd, New Delhi
5. *'Indian Deep Space Network - Mars not too Far'* by G.R. Hathwar – Publisher Notion Press, Chennai
6. DARS 2009- Proceedings of DARS 2009 PART1 and 2 - ISRO Publication

* * *

Manohar Sonnada is an expert in reliability and quality assurance as well as test and evaluation of RF systems covering TTC, remote sensing payload data handling and communication satellite payloads. He is also a specialist in the qualification of satellite control earth stations and frequency coordination aspects.

G. Raghavendra Hathwar has vast experience in establishing satellite control earth stations for LEO, GSO and Deep space missions of ISRO satellites and tracking stations and down ranging stations for launch vehicles of ISRO. He was the Associate Project Director for realizing the first state of the art deep space network system project for supporting Moon and Mars missions of ISRO. He has authored three books on satellite and earth station technology.

ATTITUDE DETERMINATION AND CONTROL SYSTEM

P. Natarajan & C.S. Prasad

Attitude determination and control system (ADCS) along with onboard computer (OBC) and digital electronics are the real brain of any satellite. Major functions of ADCS are to determine the satellite's three axes coordinates in inertial space, orientation, control and maintaining the spacecraft in the desired orientation for the payload operation.

For remote sensing satellites in low earth orbit, the payload must always be pointed towards the earth for imaging purposes. For communication satellites in geostationary orbit, the antenna must always be oriented to the earth for the required services. Whereas for scientific applications, the telescope or the probe must be continuously pointed towards that particular star or planet.

ADCS design is very challenging and tricky since it has to integrate the control algorithms with the OBC as well as digital electronics and translate the same into precession movements of the satellite body through the actuators such as the propulsion *thruster and wheels for attaining three axes stabilisation and control. Also, the ADCS has to respond to contingency operational requirements in case of loss of attitude and bring the satellite back to nominal orientation for resuming regular operations.*

The reader will get a good insight into the theory and applications of control dynamics for the satellite operations in this chapter.

The attitude determination and control system of a Nanosatellite is responsible for orientation, control and stabilisation of the satellite in the desired direction as dictated by the mission/payload requirement. The attitude of a satellite is its orientation in three-dimensional space with respect to a specified reference frame.

For communication satellites in geostationary orbit, the antenna must always be oriented towards a particular location on the earth for the required telecommunication, TV/radio broadcast or meteorological services. As the earth keeps spinning about its polar axis, the satellite also has to rotate in an axis perpendicular to the orbital equatorial plane so as to constantly point the antenna to the same location.

In the case of earth observation remote sensing satellites, the payload must be pointed continuously towards the required area of interest on the earth for imaging. In case of scientific applications such as observing stars or planets, the telescope must be continuously pointed towards that particular star/planet.

The satellite orbit keeps drifting due to various disturbances acting on it such as the earth's gravitational force and solar wind pressure. For maintaining the nominal orbital altitude and inclination, thrusters of the propulsion system are to be fired in the required direction. This is achieved by the satellite ADCS, which provides attitude information and maintains the required orientation throughout its operational life, starting from its injection into the orbit. The ADCS includes attitude measurement sensors, actuators, onboard computer (OBC), algorithms and software as well as ground support equipment used to determine and control the attitude (rotational orientation) of a satellite.

10.1 ATTITUDE REPRESENTATION

Three parameters are required to define the attitude of a rigid body in a three-dimensional space. There are various methods for the mathematical representation of a rigid body's attitude transformation or rotation. The most common ones are – direction cosine matrices (DCM), Euler angles and quaternion. Each method has its advantages and drawbacks.

10.1.1 Reference frames

Attitude is always referred to with respect to some reference frame. Several reference frames in three dimensions are of special interest for attitude analysis. In general, a reference frame is specified by the location of its origin and the orientation of its coordinate axes. **Fig. 10.1** shows the reference frames commonly used in case of satellites.

10.1.2 Satellite body frame

A satellite's body frame is defined by an origin at a specified point (usually centre of mass) in the satellite body and three Cartesian axes. A body frame is used to align various sub-systems like payload camera and attitude sensors of the satellite during satellite assembly.

10.1.3 Inertial reference frames

The earth-centred inertial reference (ECI) frame: The origin of the axes in this coordinate reference frame is fixed at the centre of the earth.

- The X-axis is parallel to the line of nodes and is positive in the vernal equinox direction.
- The Z-axis is parallel to the earth's geographic north-south axis and pointing north.
- The Y-axis satisfies the right-handed orthogonal triad.

10.1.4 Earth-centred/earth-fixed frame (ECEF)

The earth-centred/earth-fixed frame is similar to the ECI frame with the z-axis along the earth's north pole. However, the x-axis points in the direction of the earth's prime meridian, and the y-axis completes the right-handed system. Unlike the ECI frame, the ECEF frame rotates with the earth.

10.1.5 Local horizontal/ local vertical frame (orbit reference frame)

In this frame the z-axis is along the local vertical, x-axis along the local horizontal and y-axis is along the orbit normal. The origin is fixed at the centre of the mass of the satellite.

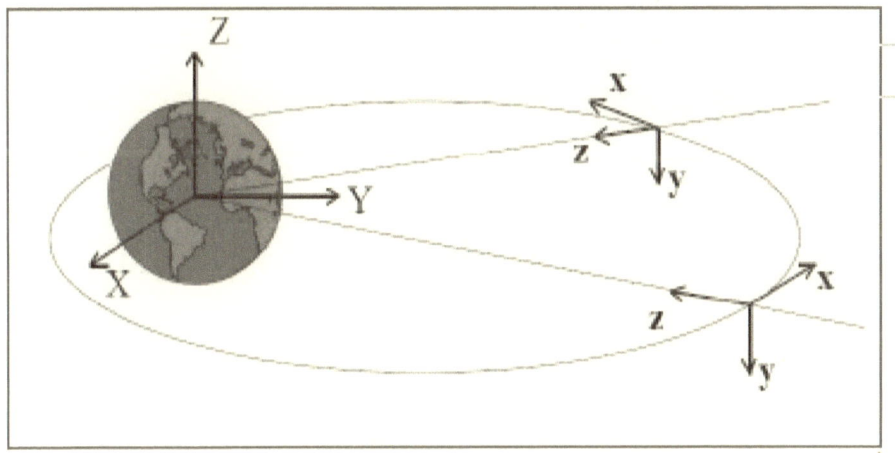

▲ **Fig. 10.1:** Reference frames

XYZ - Inertial reference frame
xyz – Orbit reference frame

10.1.6 Direction cosine matrix

A direction cosine matrix is a transformation matrix which is composed of the direction cosine values between the initial coordinate system and the target coordinate system. The direction cosine matrix transforms a vector from one coordinate system to another coordinate system

10.1.7 Euler angles

The orientation of a rigid body with respect to an inertial coordinate system can be described by three successive transformations about the body-fixed axis. The three angles used for the successive transformation are the Euler angles.

Usually, they are used for the graphical display of the satellite orientation since they are relatively easy to interpret. A body-fixed axis can be used for the first transformation. The second rotation must be performed by any of the two axes not taken for the first transformation. The final transformation is about any axes not employed by the second transformation. Therefore 12 different transformation sequences (sets of Euler angles) exist for this scheme to describe the attitude of a rigid body. The transformation matrix for the transformation sequences is obtained by the multiplication of three elementary transformation matrices.

Euler rotation sequence 3-1-3 (**Fig. 10.2.**)

XYZ – Reference frame xyz – Body frame

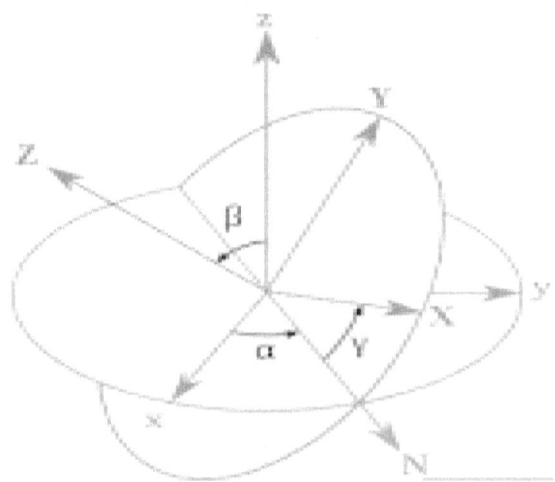

▲ **Fig. 10.2:** Euler rotation

- The first rotation is about z-axis through an angle 'α'
- The second rotation is about rotated x-axis through an angle 'β'
- The third rotation is about rotated z-axis through an angle 'γ'
- α, β and γ are Euler angles

10.1.8 Definition of roll, pitch, yaw axes

For all practical purposes, any satellite is represented by its body axes called roll, pitch and yaw axes (the same nomenclature as followed by the aerospace industry). The roll axis points along the velocity vector of the satellite in the orbit, the pitch axis is perpendicular to the roll axis and the orbital plane and the yaw axis follows the right-handed triad as shown in **Fig. 10.3(a)**. Satellite attitude errors are defined as rotations about each of these axes, that is

- Yaw error – Rotation about the yaw axis
- Roll error – Rotation about the roll axis
- Pitch error – Rotation about the pitch axis

These attitude errors measured by various sensors are also referred to in these three axes with a sign convention represented by the right-hand thumb rule as shown in **Fig. 10.3(b)** (when the right-hand thumb is pointed along the respective axis i.e., R, P, Y, the direction of the rotation of four fingers represent 'positive error' and opposite is 'negative error).

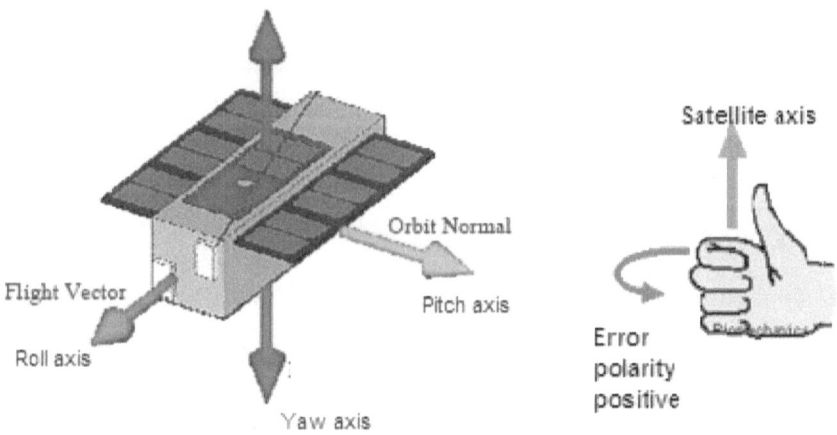

▲ **Fig. 10.3:** (a) axes definition (b) error polarity

Typical orientation of satellite axes in its orbit is shown in **Fig. 10.4.**

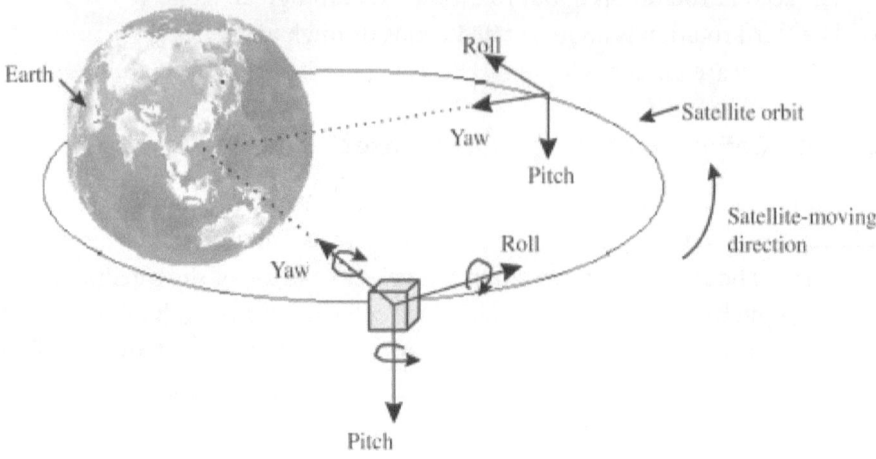

▲ **Fig. 10.4:** Satellite attitude axes

10.1.9 Quaternion

The representation of relative orientation using Euler angles is easy to develop and visualise but computationally intense. Also, a singularity problem occurs when describing attitude kinematics in terms of Euler angles and therefore it is not an effective method for satellite attitude dynamics. The widely used quaternion representation is based on Euler's rotational theorem, which states that the relative orientation of two coordinate systems can be described by only one rotation about a fixed axis. A Quaternion is a 4×1 matrix whose elements consist of a scalar part **s** and a vector part **v**.

10.2 EQUATIONS OF MOTION

A satellite's attitude changes according to the fundamental equations of motion for rotational dynamics; the Euler equation expressed in vector form in the satellite's reference frame:

$$H = T - \omega \times H$$

This vector equation represents the conservation equations for the physical vector quantity of a body or collection of bodies called angular momentum, which is denoted by H. This is similar to the linear momentum related to translational motion of a body that will remain constant unless a force acts to change it and it is calculated as mass multiplied by velocity (F = m x a). Analogously, angular momentum is the rotational motion of a body that will continue unless changed by torque and it is calculated as the body's moment of inertia multiplied by times its angular velocity. The moment of inertia is a 3-by-3 matrix of values that describe the distribution of mass in a body. There is always a coordinate frame, called the principal axis frame for which the moment of inertia matrix is diagonal.

The total angular momentum of the system is given by

$$H = I\omega + h$$

where I is the moment of inertia, ω is angular velocity and h is the angular momentum stored by any rotating object that is part of the satellite, such as momentum/reaction wheels or gyroscopes. So, by the product rule of calculus the Euler equations can be rewritten as a matrix equation:

$$I\dot{\omega} = T - \dot{h} - \dot{I}\omega - \omega \times H$$

The above equation allows us to understand how attitude changes due to various causes. The first term on the right-hand side represents the direct contribution of external torque to attitude dynamics; this term includes how some actuators can be used to control a satellite's attitude by creating external torques. The second term gives the relationship between changes in onboard rotating objects' speeds and changes in the satellite's rotational velocity; this term is where certain other control actuators like momentum wheel/reaction wheel/control moment gyro enter into the dynamics as so-called internal torques. The third term shows how changes in the satellite moment of inertia (representing how mass is distributed in the satellite), such as by solar array articulation, can affect attitude dynamics; in the absence of changes in mass properties, the third term disappears. The fourth term is called the gyroscopic torque and it shows how the angular momentum appears to change direction but not the magnitude in the satellite's frame of reference when the satellite is rotating. All these effects combine to determine the rate of change of the angular velocity on the left-hand side.

Attitude kinematics of the satellite in terms of a quaternion can be written as

$$\dot{q} = \frac{1}{2}(\Omega)(q)$$

Where,

$$(\Omega) = \begin{bmatrix} 0 & \omega_z & -\omega_y & \omega_x \\ -\omega_z & 0 & \omega_x & \omega_y \\ \omega_y & -\omega_x & 0 & \omega_z \\ -\omega_x & -\omega_y & -\omega_z & 0 \end{bmatrix}$$

10.3 ATTITUDE SENSORS

Attitude measurement/determination is a critical aspect for a satellite's mission and a wide variety of sensors and estimation algorithms are available for use in an attitude determination system. The traditional types of sensors that can be used for attitude measurement are sun sensors, star trackers, earth horizon sensors, magnetometers and rate gyroscopes. Except for the gyroscope, which is an inertial sensor, other sensors measure the attitude error of a satellite with reference to some celestial object like the sun, stars or the earth. A brief description of each type of sensor is given below.

10.3.1 Sun sensors

Sun sensors provide a measurement of one or two angles between the sensor's boresight direction and the sun, providing information about the line-of-sight vector to the sun in the body-fixed frame. A typical sun sensor consists of a pinhole or a narrow slit and a quadrant photodiode detector. The sunlight falls on the detector through the pinhole or slit and based on the position of the light beam falling on the photodiode, the pitch and roll errors are computed. Various types of sun sensors exist, namely analogue sun sensors and digital sun sensors with accuracies ranging from better than 0.1degree to a few degrees. The working principle of a sensor and a typical model is shown in **Fig. 10.5**.

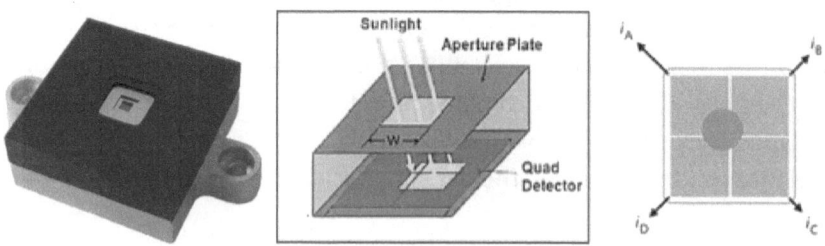

▲ **Fig. 10.5:** Typical sun sensor

10.3.2 Star trackers

Star trackers are the most accurate type of satellite attitude sensors. Most modern star trackers work much like digital cameras in that they take a snapshot of the star pattern. This star image is compared with the star catalogue stored onboard. First, the stars are identified using a pattern-matching algorithm. The reference line of sight vectors are obtained from the onboard star catalogue and measurement vectors are constructed using the measured stars. Then the full attitude solution can be derived by comparing the reference vectors and measurement vectors. Star tracker accuracies range from one arc-second to few arc-seconds depending on the quality of the sensor. A typical star tracker is shown in **Fig. 10.6** (Courtesy: Alexander O. Erlank, *"Development of CubeStar: a CubeSat-compatible star tracker"* [2013])

A star tracker consists of an imaging lens, an area array charge-coupled device (CCD) detector and processing electronics. The lens focuses stars on to the detector and depending on the magnitude of a star set for detection (typically 6th magnitude), bright stars are identified and compared with onboard catalogue stars. Since the star tracker detects low-level light from stars, a proper optical baffle needs to be used in front of the lens to minimise stray light falling on the detector. For Nanosatellites, miniature star sensors are commercially available.

▲ **Fig. 10.6:** Star tracker with baffle and electronics

10.3.3 Earth horizon sensors

Horizon sensors, also known as earth sensors, are infrared sensors that detect the radiance contrast between the cold space and the carbon dioxide emission of the earth's atmosphere in 14 to 16-micron thermal IR band. A scanning horizon sensor consists of a germanium wedge prism rotated by a DC motor that makes a conical scanning by crossing the earth's horizon at two points as shown in **Fig. 10.7**. The incoming IR radiation is imaged by a germanium lens onto a thermistor bolometer detector. Two electrical pulses generated at space to earth and earth to space crossings are sensed and processed to derive a satellite's roll and pitch errors. The chord width between the two pulses gives the roll error and phase shift between the two pulses with reference to the centre of earth gives the pitch error. Earth horizon sensors typically give accuracies between 0.1 to 0.25 degree.

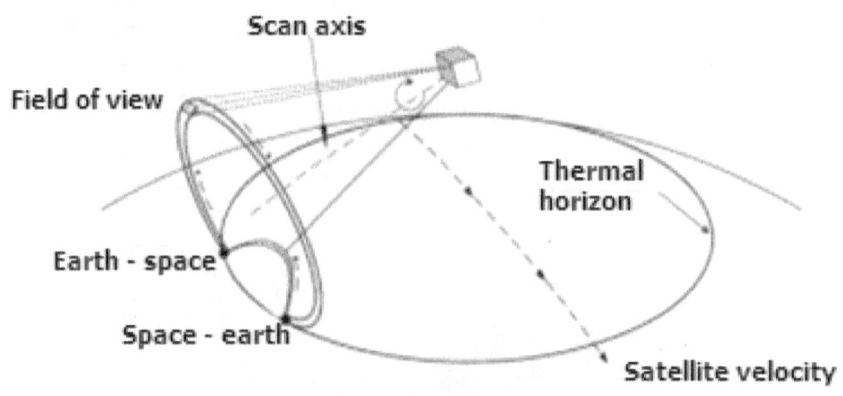

▲ **Fig. 10.7:** Principle of conical scanning earth sensor
(Courtesy: https://ars.els-cdn.com/content/image/1-s2.0-S0030402615017015-gr1.jpg)

10.3.4 Magnetometers

Magnetometers measure the direction and magnitude of the earth's magnetic field. The strength of the earth's magnetic field decreases proportionally to the inverse cube of the distance of the satellite from the earth. Hence magnetometer-based attitude determination is generally used for altitudes below 1,000 kms. A magnetometer output depends on the satellite's attitude. Reference magnetic vectors are constructed using the international geomagnetic reference field (IGRF) model. Measured magnetic vectors are compared with the reference vectors to obtain the satellite's attitude. Magnetometers operate with peaking flux gate technique where three magnetic coils mounted along the three axes of the satellite are excited with AC current to saturate/de-saturate the core. In the presence of any external magnetic field, the current peaks shift positive or negative which is a measure of the external field. For Nanosatellites, miniature magnetometers using the anisotropic magneto-resistance (AMR) principle (electrical resistance of a magnetic material changes with external magnetic field environment) are deployed. Satellite magnetometer attitude measurement accuracies typically range from 0.5 to 3 degrees.

10.3.5 Rate gyroscopes

Rate gyroscopes, also referred to simply as gyros, provide a measurement of the angular velocity of the satellite in all the three axes and integrating these rates over a time interval gives attitude angles. A typical mechanical gyroscope consists of a wheel spinning at about 6,000 rpm mounted on gimbals. Following the 'gyroscopic stiffness' principle, any torque applied to this rotating system will tilt the spin axis into a plane perpendicular to the plane of the torque applied as shown in **Fig. 10.8**. A satellite gyro will have three such wheels mounted along three orthogonal axes and each wheel provides rate data about two axes (other than its spin axis). However, mechanical gyros will have inherent drift due to imbalances and external reference sensors are used to periodically reset the drift. Gyroscopes used in Nanosatellites are configured with micro-electro-mechanical (MEMs) devices on a single silicon wafer. Fusing the gyro data with attitude measurements provides more accurate attitude estimates than the attitude measurements alone with the degree of improvement dependent on a number of factors such as the sensor sampling frequency, satellite dynamics and gyro noise characteristics.

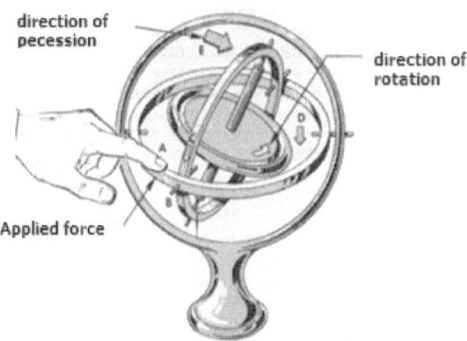

▲ **Fig. 10.8:** Principle of Gyroscope (Courtesy: https://maritime.org/)

10.4 ACTUATORS

Actuators used for satellite attitude control can be classified into three categories:
- Momentum exchange devices
- Magnetic actuators
- Mass expulsion devices

10.4.1 Momentum exchange devices

The use of momentum exchange devices for attitude control is an attractive option because they do not require any expendable fuel resources. Momentum exchange devices operate on the principle of conservation of angular momentum in a torque-free environment. External torques change the total angular momentum and internal torques exchange momentum between different rotating parts of a satellite. In this way, reaction wheels or control moment gyroscopes may be used to change satellite pointing without affecting total angular momentum. Because environmental disturbances create external torques on the satellite, they also create angular momentum that must be either stored or removed by the attitude system.

Small external torques that vary over the course of an orbit but have a mean of zero (cyclic torques) may be managed just through storage but those torques that have a non-zero mean (secular torques) will cause a gradual increase in angular momentum and this momentum build-up must eventually be removed with actuators that create external torques. Thrusters, magnetic torquers or even solar trim tabs can be used to create controlled external torques on the satellite, thus controlling the total angular momentum.

Reaction wheels are small disks attached to the satellite and driven by electric motors. The basic operation of a reaction wheel is fairly intuitive. Spinning a wheel mounted on a particular axis in the clockwise direction will make the satellite to spin in the counter-clockwise direction about the same axis so that the total angular momentum is conserved.

10.4.2 Magnetic actuation

Magnetic Torquer is often considered for small satellite attitude control because of potentially lower cost, weight and greater reliability than other actuators. In magnetic actuation, magnetic torquers are energised to produce a magnetic field, which interacts with the local earth's magnetic field to produce a torque which is used for attitude control.

10.4.3 Mass expulsion devices

Reaction jets or thrusters produce thrust by expelling mass. They may be hot gas or cold gas thrusters. Hot gas thrusters use monopropellant or bipropellant fuels. In the case of monopropellant, generally, hydrazine is used. Bipropellant uses the combination of monomethylhydrazine (MMH) as fuel and nitrogen tetroxide (N_2O_4) as an oxidizer. Cold gas thrusters produce relatively small forces by expelling a compressed neutral gas like Nitrogen.

10.5 ATTITUDE STABILISATION TECHNIQUES

Attitude stabilisation techniques of a satellite can be broadly classified into two categories: Passive stabilisation and active stabilisation. Passive control techniques take advantage of basic physical principles and/ or naturally occurring forces by designing the satellite to enhance the effect of one force while reducing the effect of others. Active control systems directly sense satellite attitude errors and supply a torque command to alter it as required. This is the basic concept of feedback control. The main techniques which constitute passive stabilisation are:
1. Gravity gradient stabilisation
2. Spin stabilisation
3. Passive magnetic stabilisation
4. Aerodynamic stabilisation

The active stabilisation techniques are:
1. Reaction wheel-based attitude control
2. Momentum wheel-based attitude control
3. Reaction control thruster-based attitude control
4. Active magnetic stabilisation

10.5.1. Passive stabilisation techniques

10.5.1.1 Gravity gradient stabilisation

The gravity gradient phenomenon can be used to stabilise a satellite in a nadir pointing attitude. Gravitational force is inversely proportional to the square of the separation distance between the satellite and the earth. In orbit, the differences in the earth's gravitational pull across the satellite mass due to the minor difference in the distance to the earth become significant sources of torques. For cylindrically shaped satellites, the length of the satellite (axis of minimum moment of inertia) will tend to align with the nadir vector. Gravity gradient stabilisation provides nadir-pointing stabilisation acting in pitch and roll to maintain a nadir-pointing attitude while leaving the yaw uncontrolled. To control this third degree of freedom, a small constant-speed momentum wheel is sometimes added along the intended pitch axis. The momentum-biased wheel will be most stable when perpendicular to the nadir and velocity vectors and therefore parallel to the normal orbit. The stable state of the gravity gradient plus momentum bias wheel establishes the desired attitude through small energy dissipation onboard without the need for active control. Frequently dampers are added to gravity gradient satellites to reduce librations - small oscillations of the nadir vector caused by other environmental disturbance.

10.5.1.2 Spin control stabilisation

Spin stabilisation is a passive control technique in which the entire satellite rotates so that its angular momentum vector remains approximately fixed in inertial space. Spin-stabilised satellites (or spinners) employ the gyroscopic stability to passively resist disturbance torques in two axes. Additionally, the spinners are generally designed to be either insensitive to disturbances around the third axis (the spin axis) or else have active means of correcting these disturbances. A satellite is stable (in its minimum energy state) if it is spinning about the principal axis with the largest moment of inertia. Energy dissipation mechanisms onboard, such as propellant slosh and structural damping will cause any vehicle to progress toward this state if uncontrolled. So, disk-shaped satellites are passively stable whereas pencil-shaped satellites are not. The principal disadvantage of spin stabilisation is that the mass properties of the satellite must be carefully managed during design and assembly to ensure the desired spin direction and stability. One thumb rule for stability is that the ratio between the maximum to the minimum moment of inertia of the satellite should be greater than 1:1. Often along with spin stabilisation, magnetic torquers are used for spin rate control and axis orientation control

10.5.1.3 Aerodynamic stabilisation

The atmospheric density decreases exponentially with altitude. For LEO orbits around 500 kms, the atmosphere is sufficient enough to drag satellites causing increased orbital decay and angular moments. Aerodynamic force can be used to provide stability aligning the required axis of the satellite with the velocity vector. Aerodynamic stability typically acts in pitch and yaw to maintain a ram-facing attitude while leaving the roll uncontrolled.

10.5.1.4 Passive magnetic stabilisation

Passive magnetic stabilisation is achieved using a set of permanent magnets onboard a satellite in the LEO to align the satellite with the earth's magnetic field lines that it experiences in the orbit. The attitude of a magnetically stabilised satellite depends upon the type of orbit. In a low inclination orbit, the magnets will tend to point towards the magnetic north like a compass needle, whereas in a higher inclination orbit such as polar orbit, a magnetically stabilised satellite would perform two cycles per orbit, where it would line up north to south over the equator and tumble over the earth's magnetic poles to line up with the earth's magnetic dipole.

10.5.2 Active stabilisation techniques

Satellites actively stabilised in all three axes are much more common today than those using spin or gravity gradient stabilisation. They can manoeuvre relatively easily and can be more stable as well as accurate depending on their sensors and actuators, than mere passive stabilisation techniques. However, they are more expensive and also often more complex but the processor and reliability improvements have allowed comparable or better total reliability than passive systems. The control torques about the axes of three axes systems are based on combinations of momentum wheels, reaction wheels, control moment gyros, thrusters, solar or aerodynamic control surfaces (e.g. tabs) or magnetic torquers. Broadly these active systems take two forms – one uses momentum bias by placing a momentum wheel along the pitch axis and the other is called zero momentum and does not use momentum bias at all - any momentum bias effects are generally regarded as disturbances. Either option needs some method of angular momentum management such as thrusters or magnetic torquers in addition to the primary attitude actuators.

10.5.2.1 Reaction wheel-based attitude control (zero momentum system)

In a zero-momentum system, actuators such as reaction wheels respond to disturbances on the satellite. For example, an attitude error in the satellite measured by a sensor results in a control signal that torques the wheel creating a reaction torque in the satellite that corrects the attitude error. This reaction torque on the wheel either increases or decreases its speed. The aggregate effect is that all disturbance torques are absorbed over time by the reaction wheels thereby causing their speeds to reach set limits. In such cases, their accumulated angular momentum needs to be removed to bring the speed within operable limits. This momentum removal, called de-saturation or momentum dumping or momentum management will be accomplished by thrusters (propulsion system) or magnetic torquer coils.

Three reaction wheels with each rotational axis parallel to the satellite axis (R, P, Y) constitute the simplest control system. As the dynamics of the three-body axes are separated, the control system design for each axis can be done separately. Redundancy is an important part of any space mission. It is, therefore, necessary to provide for the possibility of a reaction wheel failure in space. An extra wheel is, therefore, added to the system to provide continual three axes controllability even if one reaction wheel fails. The fourth wheel is mounted in the satellite with its rotational axis equally inclined to the principal axes giving equal torque component about each axis. Thus the failure of any one of the wheels along the principal axes can be compensated by the fourth wheel. The orthogonal reaction wheel configuration is shown in **Fig. 10.9**

Attitude control can be achieved using three orthogonal wheels in 3RW configuration or all the four wheels in 4RW configuration. In 3RW configuration, nominally all the wheels are operated around zero speed and hence this is also called zero momentum system. To achieve zero momentum system in 4RW configuration, the three orthogonal wheels are operated around a non-zero speed (typically around 1000 rpm) in the clockwise direction and the skew wheel is operated around 1732 rpm in the counter-clockwise direction. The component of angular momentum of the skew wheel is exactly compensated by the momentum of the orthogonal wheels because the skew rotates in the opposite direction to that of orthogonal wheels. As the nominal speed of all the reaction wheels in 4RW configuration is non-zero rpm, the effect of stiction or static friction is avoided.

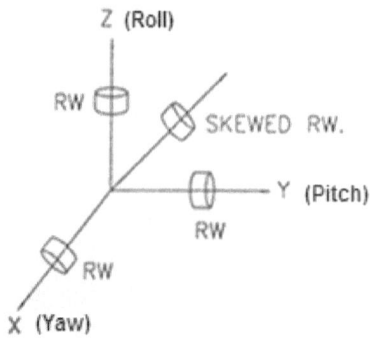

▲ **Fig. 10.9:** Four reaction wheel system

Alternately, the four reaction wheels can be mounted in a tetrahedral configuration as shown in **Fig. 10.10**, which has several advantages. Each wheel generates torque and angular momentum components along all the satellite principal axes. In a typical mounting where all the four wheels are equally spaced between the principal axes, the torque in a single axis is 2.1 times the individual wheel torques. Similarly angular momentum storage capacity also increases by the same factor. Another advantage is increased torque requirement about a particular axis can be achieved by mounting all the wheels closer to that axis. This configuration also provides a zero momentum system by operating the two wheels in a clockwise rotation and the other two wheels in the counter-clockwise rotation. This configuration is particularly useful in remote sensing satellites where rapid attitude manoeuvres are involved. The zero momentum system provides tighter control and allows greater agility.

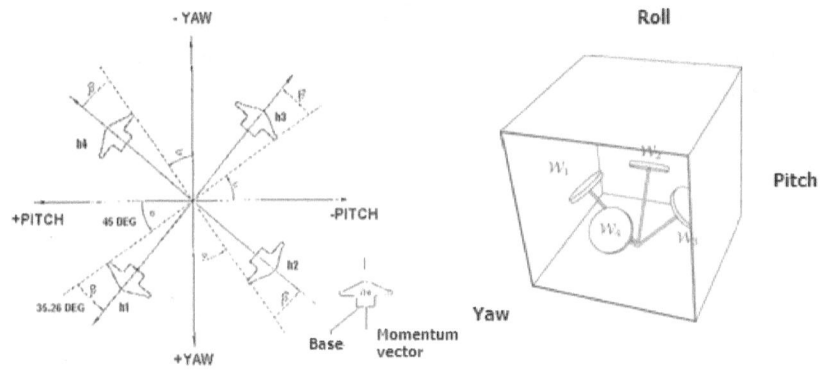

▲ **Fig. 10.10:** Tetrahedral wheel configuration

10.5.2.2 Momentum wheel-based attitude control

Momentum biased systems often have just one wheel with its spin axis mounted along the pitch axis, ideally normal to the orbit plane. The wheel is run at a nearly constant, high speed to provide gyroscopic stiffness to the satellite, just as in spin stabilisation with similar nutation dynamics. Around the pitch axis, the satellite can control attitude by torqueing the wheel, slightly increasing or decreasing its speed. Periodically, the momentum in the pitch wheel must be managed (i.e. brought back to its nominal speed) as in zero momentum system, using thrusters, magnets or other means. Reorienting the attitude of biased momentum system requires precession of a momentum vector, which can be a slow process. The bias momentum is commonly used in communication satellites because of its inherent attitude stability and does not require attitude manoeuvre except during orbital station-keeping.

Earlier satellites used a single momentum wheel along with magnetic torquers for three axes attitude control. The variability of the momentum wheel speed with respect to the bias speed provides attitude control for pitch axis. Magnetic torquers provide attitude control along the yaw and roll axes. Current satellites use two momentum wheels mounted at a small angle to the pitch axis in the pitch-yaw plane to provide the biased momentum. **(Fig. 10.11).** Modulation of the wheel speeds in the same direction provides pitch attitude control whereas differential modulation of speeds provide yaw torque which is used for roll attitude control. In the case of bias momentum system, the roll attitude control is achieved by yaw torque as per the law of precession of angular momentum, which states that the bias momentum tends to align with the torqueing axis. A reaction wheel along the yaw axis establishes redundancy for the system. Nominally two momentum wheels are operated in the V mode for two axes control with roll and pitch attitude sensed by the earth sensor. Yaw control is achieved through roll–yaw coupling because of the gyroscopic stiffness. In case of failure of one of the momentum wheels, yaw reaction wheel or the rest of the momentum wheel is operated in L mode with appropriate modulation of wheel speeds to provide roll and pitch attitude control. The magnetic torquers supplement the roll/yaw control.

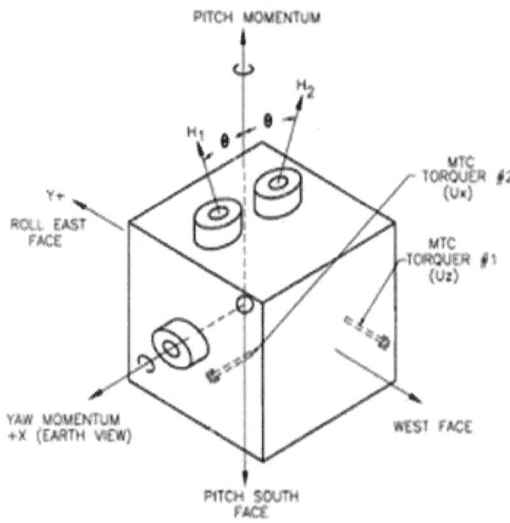

▲ **Fig. 10.11:** Momentum wheel configuration

10.5.3.3   Active Magnetic Attitude Stabilisation

A very specialised form of zero momentum control called active magnetic control can be attained from a combination of a magnetometer, a global positioning system (GPS) receiver and computationally intensive software filtering. The GPS feeds the satellite location to the onboard processor, which then determines the local magnetic field based on onboard models. The magnetometer data is filtered using Euler equations to determine the attitude and magnetic torquers make corrections in the two available directions at any given moment - corrections on the magnetic field vector are not possible. Magnetic control systems are relatively reliable, lightweight and energy-efficient and therefore magnetic attitude control is attractive for small, inexpensive satellites. They can be used as a backup control mode for a LEO satellite but can be the primary control mode for a satellite in a highly inclined orbit. (The highly inclined orbit has large changes in magnetic field direction allowing the filtering algorithm to better determine a three axes attitude solution.) This attitude knowledge can also be combined with other sensors, such as sun sensors, for more accuracy.

The operating principle is very simple and is based on the following equation

$$T = m \times B$$

where 'm' is the dipole moment of the torquer rods, 'B' is the earth's magnetic field and 'T' is the torque produced which is used for attitude control.

The torque produced is perpendicular to the earth's magnetic field. A rigid satellite has three rotational degrees of freedom but the torque rods can only torque the two axes that are perpendicular to the magnetic field vector. The system is controllable if the orbit is inclined because the earth's magnetic field vector rotates in space as the satellite moves around its orbit.

A lot of research has been done in the area of magnetic stabilisation based on the assumption that the geomagnetic field is periodic, the satellite motion is periodic and the magnetic field at any instant is known from magnetometer measurements. Often magnetic stabilisation has been used in conjunction with other stabilisation such as gravity gradient stabilisation for nadir pointing satellites. The control scheme can be simplified by actively controlling one axis using a reaction wheel and controlling the other two axes by magnetic stabilisation.

10.5.3.4 Attitude control using reaction control system thrusters

Satellites have small rocket engines (called thrusters) for orbit manoeuvring and attitude control. These thrusters produce thrust using the reaction force by exhausting gas. For supplying propellants to thrusters, propellant tanks, valves and heat control materials are necessary, which are assembled on the satellite to form the reaction control system (RCS). A detailed description of the reaction control system for small satellites is presented in **Annexure-A10.**

Monopropellant hydrazine-based RCS is used in many LEO satellites while bipropellant RCS is used for GEO satellites. Bipropellant thrusters, which produce thrust by the combustion of hydrazine-based fuel (monomethylhydrazine, MMH) and an oxidizer such as nitrogen tetroxide, are used when a large amount of velocity increment (delta-V) is required for orbit raising from transfer orbit to geostationary orbit or planetary exploration. On the other hand, monopropellant thrusters, which produce thrust by decomposition of the propellant (hydrazine, N_2H_4) in combination with gaseous nitrogen to generate high-temperature gas, are used when less velocity increment is needed.

For any kind of thruster, a higher specific impulse, operability improvement/safer handling (use of lower toxicity propellants) and cost reduction are preferred. For this reason, RCS using lower toxicity propellants called "green propellants" instead of toxic hydrazine are attracting attention. Green monopropellant thrusters use a hydroxyl ammonium nitrate (HAN) based propellant.

Pulsed plasma thrusters (PPTs) are a new option for attitude control of a small satellite and may result in reduced attitude control system (ACS) mass and cost. The PPT system can be used for disturbance torque compensation and slew manoeuvres for sun acquisition for which the small impulse bit and high specific impulse offer unique advantages. Analysis of the use of PPTs for ACS showed that the replacement of the standard momentum wheels and torque rods with a PPT system to perform the attitude control manoeuvres on a small LEO satellite reduced the ACS mass by 50 to 75 per cent with no increase in required power level over comparable wheel-based systems though rapid slewing power requirements may be an issue.

10.6 DESIGN CONSIDERATIONS FOR NANOSATELLITES

The small size of a Nanosatellite imposes severe constraints on mass, volume and power. Therefore novel designs need to be investigated and are often necessary to meet the requirements of a Nanosatellite. Experiments have been conducted using actuators such as reaction wheels, magnetic torque coils, and micro-thrusters. Active control actuators, in general, are very well understood and are widely used. The miniaturisation of these actuators remains a challenge, especially momentum storage devices, to comply with the Nanosatellite form factor and to conform to the strict mass and power budgets. Passive methods such as passive magnetic stabilisation, aero stabilisation and gravity gradient stabilisation are robust, include no moving parts, require little to no power and are attractive options for several applications.

Nanosatellites are currently being used in various mission/payload regimes that require high accuracy pointing capabilities. For example, three axes stabilised Nanosatellites are used to accurately point their payload instruments to specific objects or regions of space. Also, Nanosatellites with communication payloads, especially those employing multiple narrow-beam antennae, require tight pointing accuracies to ensure adequate antenna gain. Attitude control system technologies developed for larger satellites are being utilised to meet increasingly stringent pointing requirements of the order of 1 degree or less.

10.6.1. Factors affecting ACS configuration

Selection of an ACS configuration and the sizing of its components is a complex function of various parameters that are typically at odds with one and other. Some of these parameters are constrained while others are variable within a certain range. The main factors which affect the selection of an ACS are the following:

1. Stabilisation requirements
2. Payload
3. Configuration
4. Accuracy requirements
5. Disturbance torques (internal and external)
6. Design constraints
7. System budgets (mass, power and volume)
8. Cost

10.6.1.1 Payload stabilisation and pointing requirements

Mission objectives are limited by available budget to determine payload instrumentation, orbit selection, stabilisation and pointing requirements. These parameters set the requirements and the range of variability for the ACS and its components. Mission and system requirements dictate stabilisation parameters and place the ACS into either spin or three axes stabilisation configuration.

Spin stabilisation is generally less complex and less expensive than the three axes stabilisation and is well-suited for some scientific experiments. Three axes stabilisation, however, is required for communications, earth observation and astronomy missions. Stabilisation systems using reaction wheels as control torque sources are well-suited for Nanosatellite applications due to their relative simplicity, versatility and capability of providing high accuracy pointing control.

Magnetic torquers are also used along with reaction wheels for momentum de-saturation. Three axes stabilisation using magnetic torquers is attractive for LEO Nanosatellites where pointing accuracy requirements are not very stringent. Magnetic control systems are relatively lightweight, require low power and inexpensive.

10.6.1.2 Configuration

The mass moments of inertia of a satellite are critical to the selection and operation of the ACS. Moments of inertia affect the sizing of the actuator and effect of disturbance torques on attitude pointing and stability. Other satellite configuration parameters influence the effect of environmental disturbances. For example, surface properties affect the solar radiation pressure. The location of the centre of pressure with respect to the centre of mass, surface area and the coefficient of drag affect the aerodynamic torques in LEOs. The overall configuration of the satellite may impose design constraints on mass, power and volume.

10.6.1.3 System budgets

System budgets for mass, power and volume and the allocation for ACS define an envelope for workable options. Identifying a workable configuration in compliance with the system budget restrictions enables system engineers to work out the design margins and shortfalls.

Orbit and environmental disturbances

The orbit of a satellite and configuration determine the type of disturbances expected during the operation. The magnitude of the environmental disturbance torques is one of the several factors that affect the sizing of ACS components. The various environmental disturbance torques which act on LEO are:

1. Gravity gradient torque
2. Solar radiation torque
3. Aerodynamic torque
4. Magnetic torque

Environmental disturbances are not necessarily detrimental because they can be used for passive stabilisation by proper design of the satellite.

10.7 DESIGN OF A TYPICAL ADCS FOR A NANOSATELLITE

The attitude determination and control system (ADCS) plays an indispensable part in a satellite's on-orbit operation that could greatly affect the satellite's performance. The development of a Nanosatellite requires an attitude control system that is inexpensive, light-weight with small volume and low power consumption. Therefore magnetic coils and bias momentum wheel have been adopted as the most popular actuators. Three axes magnetic coil combined with a pitch bias momentum wheel is a popular way to accomplish the three axes stabilisation control. The momentum wheel is installed on the pitch axis and provides gyroscopic stiffness for roll and yaw axes stabilisation. The magnetic coils are used to damp out the initial rates and also used for three axes attitude control. The measurement sensors include sun sensors, magnetometers and a three axes gyro. In most of the Nanosatellites, the power system module, onboard computer (OBC) module, the ADCS module and the telemetry telecommand (TTC) module are integrated onto a single printed circuit board (PCB). Consequently, the new single-board architecture greatly reduces the structure. Many Nanosatellite operating in orbit today adopt this control strategy. A typical ADCS structure is shown in **Fig. 10.12.**

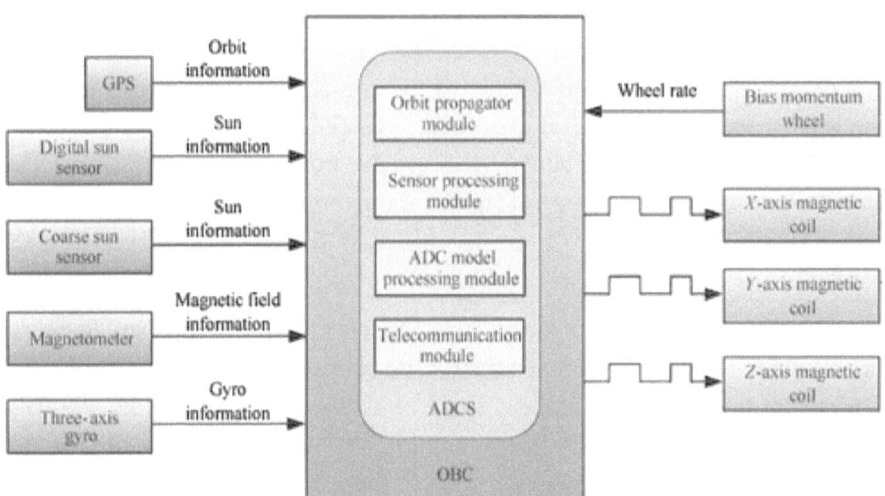

▲ **Fig. 10.12:** Typical ADCS structure

The ADCS begins to operate after the satellite separates from the launch vehicle and it is powered ON:

- After separation from the launcher, the ADCS must start automatically and get the initial attitude state
- To dump the initial angular velocity (body rates) with three axes magnetic coils using de-tumbling algorithm
- To determine satellite attitude angle and angular velocity with magnetometer and sun sensor using the Kalman filter method
- To achieve three axes nadir pointing stabilisation and maintain it throughout the mission life
- To take care of any contingencies such safe mode

Typical performance requirements of ADCS are shown in **Table. 10.1.**

▼ Table. 10.1

Sl No.	ADCS Performance	Value
1	Attitude determination accuracy (deg)	< 1
2	Attitude control accuracy (deg)	< 5
3	Attitude stability (deg/sec)	< 0.1

10.7.1 ADCS hardware architecture

Sensors and actuators are critical parts for the ADCS. Magnetometers and sun sensors are designed to obtain vector measurements. Sun sensors measure the line-of-sight vector from the satellite to the sun and magnetometers measure the local magnetic field vector. Magnetic coils generate the control torque by interacting with the earth's magnetic field and the momentum wheel stabilises the attitude through the production of angular momentum with gyroscopic stiffness.

10.7.2 SOC analogue sun sensor

A sun sensor on a chip (SSOC), shown in **Fig. 10.13,** is a two-axes and low-cost sun sensor for high accurate sun-tracking, pointing and attitude determination. The device measures the incident angle of a sunray in two orthogonal axes, providing a high sensitivity based on the geometrical dimensions of the design. A Nanosatellite's SOC sun sensor is based on the MEMS fabrication processes to achieve highly integrated sensing structures. The SOC sun sensor has a minimum size, weight and power consumption to be the perfect ADCS solution for Nanosatellite platforms.

SSOC specifications:
- Two axes measurement
- Field of view: +/- 60 deg
- Accuracy: 0.5 deg
- Size: 27 x 14 x 6 mm
- Mass: 4 gm

▲ **Fig. 10.13:** Sun sensor on chip (Courtesy: https://product.statnano.com/)

10.7.3 MEMS inertial measurement unit

Inertial measurement unit, shown in **Fig. 10.14,** consists of a gyroscope, magnetometer and accelerometer. The gyroscope measures the rotation rate about all the three axes. The magnetic component vector in the body frame is obtained directly from the magnetometers. The magnetic field is another effective vector for attitude determination, and it is the main reference vector for attitude determination in eclipse. Linear acceleration is detected by an accelerometer, which is not used in a Nanosatellite.

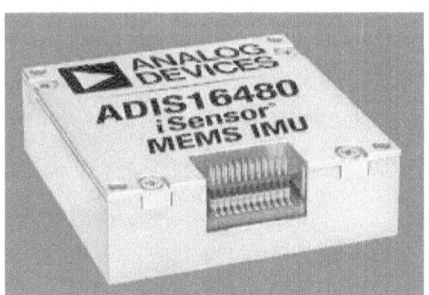

▲ **Fig. 10.14:** MEMs. Inertial Measurement Unit
(Courtesy: www.unmannedsystemstechnology.com)

Specifications

Gyroscope:
- Dynamic Range: 1000°/sec
- Bias Stability: 14.5°/hr
- Random walk: 0.66 °/√hr

Magnetometer:
- Dynamic Range: + 1.9 gauss
- Sensitivity: 145 µgauss/LSB
- **Fig. 10.10** Inertial Measurement Unit

10.7.4 Momentum wheel

A pitch bias momentum wheel is utilised as an attitude control actuator as shown in **Fig. 10.15.** In construction, the momentum wheel consists of a flywheel, bearing assembly and electric drive motor. There are also electronics for driving the wheel, controlling and measuring the wheel speed. The wheel is based on a DC micro-motor. The momentum wheel is mounted along the pitch axis, which provides the pitch axis stabilisation with a gyroscopic effect and ensures that the pitch axis point along the orbit is normal.

▲ **Fig. 10.15:** Momentum wheel
(Courtesy: https://imgbin.com/png/J6p6ELbJ/reaction-wheel-nanosat-small-satellite-cubesat-png)

▲ **Fig. 10.16:** Magnetic torquer rod
(Courtesy: https://www.cubesatshop.com/)

10.7.5 Magnetic torquer rods

A typical Magnetic torquer rod is shown in **Fig. 10.16.** Three magnetic coils are mounted along the three axes. The torque generated by the torquer rods is used for active three axes attitude control. To improve the measurement accuracy of the magnetometer, the magnetic torquer rods are operated at different time periods with the magnetometer. To reduce the residual effect of the magnetic coils, the magnetometer is mounted farther from magnetic rods. Moreover, ground tests and pre-launch calibration are introduced to improve the accuracy of the magnetometer's measurement.

10.7.6 Attitude determination

Attitude is usually expressed in terms of Euler angles or direction cosine matrix or quaternions. There are different methods for attitude determination. After a satellite separates from the launch vehicle, the attitude is unknown. Quaternion estimator (QUEST) algorithm and three axes attitude determination (TRIAD) algorithms are the most commonly used deterministic algorithms for attitude determination. Both the methods use the measurements from the sun sensor and the magnetometer. The reference sun vector and magnetic field vector are derived from the sun model and the IGRF model. Due to low computation burden, the QUEST is preferred over TRIAD. Once the three axes stabilisation is achieved, the QUEST method would provide the initial state for the Kalman filter-based algorithm and the QUEST method would be substituted by the Kalman filter method. The Kalman filter method makes use of the satellite's attitude kinematics and dynamics to estimate the states with measurements from the sun sensor and magnetometer. The accuracy of the attitude estimation depends on the sensors' accuracies.

10.7.7 Attitude control

Use of magnetic torquers to control the attitude of a satellite offers a lightweight, smooth and cost-effective method of control. However, the torque generated through the use of magnetic torquers is constrained to lie in the plane orthogonal to the local magnetic field vector with one axis being instantaneously under-actuated. If the satellite is on an inclined orbit, a suitable variation of the magnetic field allows controllability in the long-term but presents a significant challenge from an immediate control perspective. Addition of a pitch bias momentum wheel to the three magnetic torquers helps in reducing the complexity of the satellite attitude control algorithms.

10.7.8 De-tumbling control

After separation from the launch vehicle, the satellite is assumed in random initial attitude state with large angular velocity (body rates). The ADCS has three distinct control modes. The first high angular rate de-tumbling mode will serve to recover from the initial tumble conditions up to 100°/s. The B-dot method is the most popular algorithm in this phase due to its fast convergence and low computation burden. Once tumble rates have been lowered to below 30°/s, the second de-tumbling control mode will further lower the rates and place the satellite in a stable and known tumbling motion – called Y-Thomson spin. In this mode the satellite will end-up spinning only about the Y-axis and the spin axis will align itself with the orbit normal. The third control mode, Y-momentum mode, can only be activated once the satellite is in this stable tumbling state. In the Y-momentum mode, the satellite will stop spinning and stabilise to the nominal orientation (zero roll, pitch and yaw angles). In the Y- momentum mode the pitch angle may be controlled to a specific reference value using a telecommand.

Three axes stabilisation control: After the phase of de-tumbling control, the proportional-derivative (PD) control law is adopted to adjust the satellite to the desired orientation.

Tc = (-Kp*Attitude - Kd*angular rate)

Where Tc is the control torque, Kp is the proportional control gain and Kd is derivative control gain.

The required control torque is translated into dipole moment for all the three torquers accounting for only the perpendicular torques to the local magnetic field.

10.7.9 Safe control mode

The safe control mode is very important in the attitude control process. Once any ADCS component fails, attitude errors start building up and power generation also reduces. The deviation from the normal attitude is detected using the sun sensor and the satellite is automatically commanded to the safe mode to generate adequate power and also thermally safe orbit using the available ADCS components.

10.7.10 Simulations

Once the control algorithms are finalised, rigorous control analysis needs to be carried out for stability margins and robustness. The next step is to simulate the control algorithms along with the high fidelity models of attitude dynamics and kinematics, sensor models, actuators models and disturbance torque models (internal and external). These simulations are done on a desktop computer simulating all possible combinations of attitude and angular rates and for all the mission scenarios.

10.7.11   Onboard Computer

The onboard computer (OBC) acquires data from all the sensors, processes them and feeds to the control algorithms. The output of control algorithms in terms of the actuating signals are sent to the actuators such as the magnetic torquers and momentum wheel. The actuators, in turn, produce the required torque and correct the attitude. The required ADCS mode and controller/ estimator parameters are selected by telecommand. The telemetry information of all the ADCS parameters are processed by the OBC and sent to the ground for health monitoring.

10.7.12   ADCS software verification and validation

Software verification is the testing done to make sure the software code performs the functions that it has been designed to perform. Validation is the testing performed to prove that the verified code meets all of its requirements in the target system or platform. At the lowest level, each software unit is tested in a process called unit testing. When all the software units have been successfully tested, they are integrated into the full code. The integrated code is tested by computer simulation to ensure that it satisfies the software architecture requirements. Once the integrated software is validated to be correct, the last step of validation is to check that the software operates correctly on the target platform. This testing is usually performed using high-fidelity computer simulations first, followed by processor-in-the-loop simulations (PIL) and hardware-in-the-loop (HIL) simulations.

REFERENCES

1. Sidi M J 1997, 'Spacecraft Dynamics and Control a Practical Engineering Approach'
2. Wertz, James R. Ed. 1978. 'Spacecraft Attitude Determination and Control. Dordrecht, The Netherlands': D. Reidel Publishing Company.

3. Wie, Bong. 2008. *'Space Vehicle Dynamics and Control'*, 2nd edition, Reston, VA: AIAA Inc.

4. Kaplan, Marshall H. 1976. *'Modern Spacecraft Dynamics and Control. New York'*: John Wiley and Sons.

5. Agrawal, Brij N. 1986. *'Design of Geosynchronous Spacecraft'*: Englewood Cliffs, NF: Prentice-Hall, Inc.

* * *

P. Natarajan is a satellite attitude control dynamics design expert who was instrumental in the design of most of the control systems of ISRO satellites. He played a key role in the design and development of on-board autonomy and other features for most successful Indian deep space missions like Mangalyaan-1 (Mars) and Chandrayaan-1 (Moon).

C.S. Prasad, a propulsion system expert, has designed and developed several propulsion systems in LPSC-ISRO for satellites and launch vehicles. His area of specialisation is design and development of liquid propulsion components and systems for satellite mono and bi-propellant technologies.

PRODUCT ASSURANCE

L.S. Satyamurthy & B.V. Prasad

Quality and reliability assurance, which are also termed as product assurance, are specific to products such as automobiles, rockets or satellites. Generally terminologies such as 'product assurance' and 'quality assurance' are interchangeably used to primarily address the robustness and integrity for successfully developing space system hardware and software.

Quality assurance methods contribute to meticulously ensuring the quality of design, quality of conformance and quality of performance. Its importance increases with the complexity, cost and risk factors of a satellite mission. Product assurance is an integrated logistic system covering safety, service availability, ground segment and human interfaces for achieving a successful mission. For a Nanosatellite project conceived by any educational institution or a start-up company, the application of product assurance methodology will be the key to the success of the mission.

Members of a quality assurance team for a satellite programme can belong to any branch of engineering discipline, including industrial and production engineering.

The details explained in this chapter cover the product assurance methodology, statistical theory for reliability analysis, software quality assurance and overall mission assurance for a Nanosatellite and will kindle real interest in the reader.

Product assurance (PA) is an important engineering discipline of the aerospace industry in general and space systems in particular. The very fact that any space system including a satellite, once launched into space, cannot be repaired (unlike ground operated systems) implies the importance of the PA's role. The primary objective of 'assurance' is to ensure that the designed, produced and delivered space products accomplish their mission objectives safely and reliably.

Product assurance functions effectively cover almost all the engineering disciplines spread across different phases of the satellite's development cycle commencing from mission definition, sub-system design, fabrication, test and evaluation, satellite integration and up to the development of the deliverable satellite for in-orbit operations.

ENVIRONMENTAL CONDITIONS A SATELLITE UNDERGOES

Before going into details of PA aspects, it is important to understand the environmental conditions that any satellite will undergo and measures to be taken to protect it. During its life cycle, a satellite is exposed to many harsh environments including the ones on the ground such as handling/corrosion/ ageing of components. During the launch phase, the satellite is subjected to high levels of acoustic noise, vibrations and transitory g-forces (during launch vehicle stage separations), potentially causing stresses in susceptible materials and assembly joints. Once in orbit, the space environment is far from welcoming. Day-night orbital passes going through eclipse and sunlit phases cause thermal cycling (-269 degrees Centigrade to + 100 degrees Centigrade), triggering abrupt temperature changes. Prolonged exposure to hard vacuum (> 10e-6 torr; a unit of pressure) affects material characteristics. Impingement of ionised radiation/charged particles (electrons, high energy protons, neutrons) severely alters the functionality of the electronic circuit and the integrity of optical surfaces.

The key to protecting a satellite is to minimise the effects of different kinds of environments encountered at different phases with a good understanding of the behaviour of used materials and manufacturing processes. System engineering and product assurance teams need to work in tandem to realistically interpret environmental effects, assess their impact, strengthen designs and evolve withstanding methods as well as quality assurance protocols.

Due to ever-increasing requirements for autonomous intelligent systems, apart from the hardware, the role of the onboard software is also becoming vital. The complexity in developing reliable and bug-free software has also increased, resulting in a strong need for its independent assessment and rigorous quality assurance.

This chapter provides different facets of quality and reliability engineering theory and practices covering basic definitions, general principles of product assurance practices and nuances of approaches for building robust Nanosatellite programmes along with aspects of software quality assurance.

11.1 PRODUCT ASSURANCE AND QUALITY ASSURANCE

For space systems, terminologies such as 'product assurance' and 'quality assurance' are frequently referred to. Both are assurance management entities, primarily aim to address the robustness and integrity of the product. But often they are interchangeably applied while treating assurance requirements for space systems.

While most of the attributes are common between the two, subtle differences are discernible in the overall context of a project management framework. Any quality assurance programme does not generate physical products such as avionics hardware, mechanical assembly or embedded software. The output from product assurance efforts are essentially recommendations of check and balance on design trade-offs, statements of risk, procurement choice, identification of potential design weaknesses and so on.

In practical terms, the essence of any quality assurance plan for any product or system or project is categorised and summarised in aerospace parlance as:
• Paying attention to details
• Handling of uncertainties

11.2 QUALITY, QUALITY ASSURANCE AND PRODUCT ASSURANCE

Quality: Defined as 'product is fit for use and conforms to customer specification economically'. A quality product is one which is free from deficiencies.

Quality assurance: Defined as a system of planned activities known in advance for developing a defect-free product. Covers quality of design, quality of conformance, quality of performance and life assurance (reliability).

Product assurance: specific to a product (LEO/GEO/Interplanetary/Nanosatellite)
This is a discipline devoted to the study, planning and implementation of activities intended to assure that the design, controls, methods and techniques of a project result in a satisfactory level of quality in the product (ECSS-P-001A).

Both are committed to quality throughout the entire supply chain of consistently producing successful space missions and in this chapter, quality assurance and product assurance are seamlessly treated as a single entity and referred to as product assurance.

11.2.1 Quality of design:

Design assurance at the design phase is to ensure meeting of functional requirements, performance requirements and to know design constraints within a satellite's configuration. It is to ensure that quality is designed into the product.

Examples:
Mechanical systems:
- Verify adequate design margins exist with respect to expected mechanical and thermal loads/stresses
- Ensure that materials used meet the space environment (non-magnetic, non-corrosive) and out-gassing/de-gassing requirements such as total weight loss (TWL) and collected volatile condensable material (CVCM)
- Ensure Life test requirements for rotating elements (motors, gyros, wheels and scanner) and one-shot operating systems such as deployment mechanisms and pyrotechnic devices

Electrical systems:
- No single point failure that can jeopardise the mission should exist
- Selection of electronic components for circuits should be in the preference order – space-grade, MIL-883B, industrial grade and commercial of the shelf (COTS) with heritage
- All components shall follow specified 'derating' guidelines
- Precautions to be taken for hardware and software design to mitigate space radiation effects
- All power lines to be protected with properly rated fuses
- Timing margin to be ensured in the cyclic operation of microprocessors
- Potential 'hot spots' to be avoided with high power dissipating devices by providing heat sinks
- To minimise electrical interference, guidelines on electrical grounding and interference requirements like EMI/EMC shall be implemented

11.2.2 Quality of conformance

Quality control is ensured during the satellite development/fabrication phase to verify that the product is built as designed. This includes conformance of materials, processes, dimensions, sub-assembly, assembly and workmanship during all phases of a satellite's construction.

The construction of a satellite's sub-systems starts by ensuring that all the processes are qualified, personnel are trained, and external vendors are certified with all process documentation available. During the construction process, in-process control with online inspections, various checkpoints and sample tests need to be identified and implemented. Any deviations observed during fabrication shall be immediately reported to concerned agency for review, suitable correction/rework and retest as applicable.

This process is termed as **non-conformance control**. Non-conformance control procedure is applicable at every phase of design, development, manufacture, satellite assembly, integration and test (AIT) until the launch of the satellite.

The emphasis is on minimising errors in the fabrication/production process as one cannot afford any in-orbit failures.

11.2.3 Quality of performance

Quality of performance is assured during the testing phase to verify the satellite performs its intended functions at ambient and specified environmental conditions.

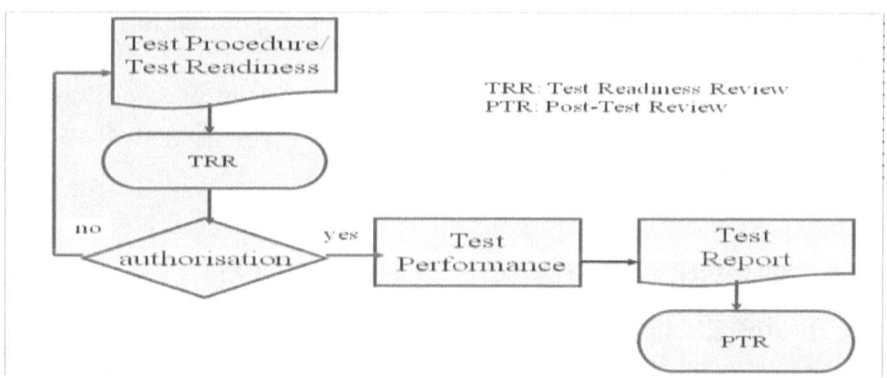

▲ **Fig. 11.1:** Typical testing flow

Test surveillance is performed by quality assurance personnel by continuously monitoring tests requiring manual intervention, periodic monitoring, test results review and validation.

The quality assurance team plays the lead role in the final acceptance of flight hardware/software systems and must be directly involved in their performance evaluation.

11.3 RELIABILITY ANALYSIS

Reliability analysis is a part of a mission's life assurance technique adopted for all space systems and it is carried out with analytical approach for assessing and ensuring that the satellite is defect-free and all potential failures are identified and corrective actions are addressed appropriately.

11.3.1 Failure mode effects analysis (FMEA)

Identification of failure modes in parts/sub-assembly/assembly/and its overall impact on the satellite performance within the sub-system configuration.

11.3.2 Failure mode effects and criticality analysis (FMECA)

Identification of failure modes within the sub-system configuration and ranking them in the order of criticality.

FMEA and FMECA are also called 'bottoms up' approach performed at the electronic component circuit level for identifying any single point failure.

11.3.3 Fault tree analysis (FTA)

Conducted at the system level where system failure is traced to lower-level events. Also called the 'top down' approach.

All these reliability analysis techniques are briefly explained in the succeeding sections for the benefit of reliability engineering students and practising faculty.

11.4 RELIABILITY STANDARDS AND SPECIFICATIONS

In the field of reliability, there are only a few yardsticks presently available. These are the International Electro-Technical Commission (IEC) specifications and the United States Military Standards (MIL-STDs). US military has published several handbooks (referred to as MIL-HDBK-XXXX), which are taken as guidelines for fabrication and testing of various satellite components/sub-systems and systems.

The British Standards Institution (BSI) has issued BS5760 – a guide on the reliability of systems, equipment and components.

Similarly, a set of Indian standards/guidelines were evolved by ISRO's Standards Organisation known as ISRO Standards for Reliability (ISREL) and are issued both for satellite and launch vehicles.

11.4.1 Failure/failure rate

Any failure is defined as the termination of the ability of an item to perform its required function. The relationship between failure and time is called as the failure rate usually represented by the symbol $\lambda(t)$. Generally, failure rates are expressed as failures per component-hours. The total number of failures within a lot of components/units, divided by the total number of life units (hours) expended by that lot during a particular measurement interval under stated conditions gives the failure rate of the component/unit.

Another way of describing the occurrence of failure is to state the meantime between successive failures. This parameter is known as the mean time between failures or MTBF and is represented by the symbol θ. MTBF is mostly used for maintainable systems, which can be in the form of periodic preventive and corrective maintenance.

11.4.2 The bathtub curve

Equipment reliability is frequently described by the familiar bathtub curve shown in **Fig. 11.2**. The bathtub curve is a composite curve formed by the addition of three separate curves or phases - the first curve represents infant mortality phase, the second curve represents useful life phase and the third curve is of wear-out phase.

During the first phase of infant mortality, weaker parts are identified and removed from the population, either by a formal screening programme or through normal equipment operation by which the instantaneous failure rate or the hazard rate decreases rapidly to a more or less residual level.

THE BATH TUB CURVE

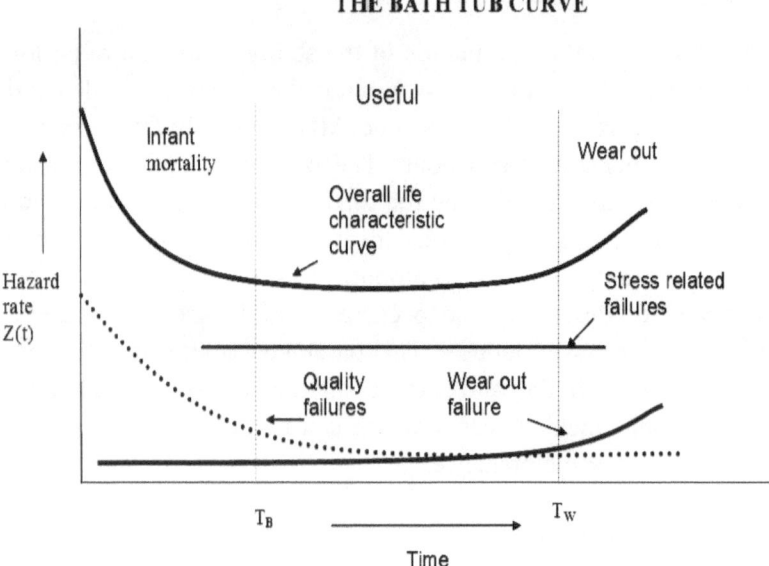

▲ **Fig. 11.2:** Bathtub curve

The second and most important phase is the useful life period. During the useful life period, failures can still occur occasionally but they are random in nature and are randomly distributed with respect to time. The statistical frequency of these failures can be predicted with fairly good accuracy but the exact time and location of actual failures cannot be pinpointed and predicted.

The third and final phase of equipment life is wear out. During this phase, the accumulated damage due to the applied stresses begins to take its toll. The components generally become weaker, more prone to failures and thus they fail with increasing frequency.

11.5 RELIABILITY PREDICTION

Reliability prediction is the process of quantitatively assessing an equipment design relative to its specified reliability requirement. During design and development, predictions serve as quantitative guides by which design alternatives can be judged relative to reliability.

The two different types of reliability predictions that must be addressed are:
- Basic reliability prediction
- Mission reliability prediction

The basic reliability prediction utilises the basic reliability model, that is, a series model. It assumes that every single part in the equipment is essential at all times and is used for estimating the demand for maintenance and logistic support.

The mission reliability prediction utilises the mission reliability model and is intended to depict the intended utilisation of the various elements within the system to achieve mission success. It takes into account redundancy, duty cycle and so on.

11.5.1 Reliability block diagrams and models

A reliability model or models of the system, sub-system or equipment is/are required for making numerical apportionments. This is essential for evaluating complex series-parallel equipment arrangements.

Basic information for the reliability model is derived from schematic block diagrams depicting functional relationships between the sub-systems and components available to provide the required performance. The reliability model re-orients the diagrams into a series-parallel network showing reliability relationships among the various components and sub-systems.

Further details on the reliability analysis including mathematical formulae are given in **Annexure A-13.**

11.6 ELECTRONIC COMPONENTS DERATING

Derating is defined as limiting the stress applied to components to levels that are well within their specified or proven capabilities in order to enhance their reliability.

Derating of all components is done with reference to their absolute maximum ratings (ex: current, voltage or temperature). These ratings are defined by the manufacturer in their specifications or datasheet as those values which should not be exceeded under any service or test condition.

Derating is effective because the failure rates of most components decrease as the applied stress levels are decreased below their rated value. The failure rate models of most electronic components are stress and temperature-dependent. MIL-HDBK-217 failure rate data shows that component failure rates vary significantly with temperature.

Derating provides an allowance for system electrical transients and also allows for possible non-uniform components heating without catastrophic failure.

Derating can help to compensate for many of the variables inherent in any design. All electronic components produced on an assembly line are not identical. Subtle differences and variations exist from one component to the next. Proper component derating will help compensate for such variations and minimise their impact upon the equipment reliability.

MIL-STD-975-M guidelines form the baseline for all the standards and have wide applicability for different missions.

11.6.1 Failure modes, effects and criticality analysis (FMECA)

The failure modes, effects and criticality analysis (FMECA) are composed of two separate analyses, the failure modes and effects analysis (FMEA) and the criticality analysis (CA).

The FMEA utilises inductive logic or a "bottom up" approach. It begins at the lowest level of the system hierarchy (for example, a small electronic component) and using knowledge of the failure modes of each component, it traces up through the system hierarchy to determine the effect that each potential failure mode will have on a system's performance. Major benefits derived from an FMEA are as follows:

- Identifies single-point failures critical to mission success or personal safety
- Early visibility of system interface problems
- A method for selecting a design with a high probability of operational success and adequate safety
- Criteria for early planning of necessary tests
- Quantitative, uniformly formatted input data for reliability prediction, assessment and safety models
- A basis for troubleshooting procedures
- A list of potential failures which can be ranked according to the seriousness of their effects and the probability of their occurrence

Criticality analysis
The criticality analysis may either be qualitative or quantitative in nature depending upon the data available and the depth of analysis desired. A quantitative approach will require specific failure data, whereas the qualitative approach will not.

The qualitative approach determines criticality on the basis of the probability of occurrence of a specific failure mode.

The quantitative criticality is determined based upon four factors:
- Probability of system loss
- Item failure mode
- Failure mode distribution ratio
- Unreliability of the item

11.6.2   Faulty tree analysis (FTA)

The fault tree is based upon deductive reasoning, that is, reasoning from the general to the specific. A specific fault is postulated and then an attempt is made to find out the modes of system or component behaviour that contributed to this failure.

The fault tree analysis focuses on one particular undesired event at a time and determines all possible causes of that event. The undesired event is the top event in that fault tree diagram. It is generally a complete or catastrophic failure rather than a drift type of failure.

Fault tree methods can be applied in the early design phase and then progressively refined and updated as the design evolves to track the probability of an undesired event.

Potential applications FTA:
- Comparison of alternative design configurations from a safety point of view
- Identification of critical fault paths and design weakness for subsequent corrective action
- Evaluation of alternative correction approaches
- Allocation of critical failure mode probabilities among lower levels of the system

FMECA and FTA are complimentary and basically equivalent methods of risk analysis. The choice between these two methods depends on the nature of the risk to be evaluated. A major advantage of the FTA is its ability to address human errors, which the FMECA cannot address.

The fault tree itself is a graphic model of the various parallel and sequential combinations of faults that will result in the occurrence of the pre-defined undesired event. The fault tree logic diagram is constructed for all possible sequences of events whose occurrence would produce the undesired events identified in the functional block diagram.

Once the fault tree has been constructed it is evaluated to obtain qualitative and/or quantitative results.

11.6.3   System redundancy concept

In reliability engineering, redundancy is defined as the existence of more than one means for accomplishing a given task. For a satellite system, redundancy is defined in terms of an alternative path of operation in case of failure of the main path. For example, if the main power supply fails, then we can select the redundant power supply.

Under certain circumstances during system design, it may become necessary to consider the use of redundancy to reduce the probability of system failure and to enhance system reliability by providing more than one functional path or operating element in areas that are critically important to mission success. However, redundancy implies increased complexity, increased weight and space as well as increased power consumption.

Types of redundancy:
(a) **Active or hot redundancy**: In this case, both systems are in the ON condition and one of them will be in operating mode. In case of its failure, the second system automatically starts operating. With this class of redundancy, external components are not required to perform a detection, decision and switching function when an element in the structure fails.
(b) **Standby or cold redundancy**: In this case, only one system will be in the ON condition and operates. If it fails, the second system is switched ON either manually or automatically. That is external elements are required with this class of redundancy to detect, make a decision and then to switch to another element or path as a replacement for the failed element or path.

Techniques related to each of these two classes are listed below:
• Simple parallel
• Series – parallel
• Duplex
• Parallel series
• Majority voting
• K-out-of-n configuration

A combination of series and parallel redundant elements may be used to protect against both short and open circuit failures. A direct short circuit across the network due to a single element shorting is prevented by redundant elements in series. An open circuit across the network is prevented by parallel elements.

The objective of duplex redundancy is to prevent incorrect logic elements from upsetting other circuits. It is used primarily in onboard computer applications.

Majority voting

A decision can be built into the basic parallel redundant model by inputting signals from parallel elements into a voter to compare each signal with the remaining signals. Valid decisions are made only if the number of useful elements exceeds the number of failed elements.

M-out-of-n configuration

In some active parallel redundant configuration, m out of n units may be required to be working for the system to function. This is called as m out of n (or m/n) parallel redundancy.

11.7 PRODUCT ASSURANCE FOR A NANOSATELLITE

For a Nanosatellite project, major guidelines for product assurance depend on the overall philosophy adopted by the institution which is fabricating the satellite, in terms of:
- Envisaged mission goals and mission life
- Model philosophy to be adopted
- Satellite development plan (develop in-house or procure from outside)
- Expertise available and training of personnel
- Resources available in the institution
- Project cost and schedule

In most of the cases, Nanosatellites developed by educational institutions are meant either for technology demonstration or a specific scientific experiment with a limited mission life of one or two years. Also, due to resource constraints, they adopt a single model (flight model) approach with a limited bread-boarding of sub-system designs. Depending on the expertise, facilities and funds available, the institution may either manufacture the Nanosatellite in-house or procure from vendors at sub-system level or even a full satellite. The product assurance plan needs to be tuned to suit these scenarios, with a single objective of ultimately developing a reliable Nanosatellite to meet the mission goals.

On the other hand, the current global scenario shows that Nanosatellites are manufactured by the industry in large numbers and deployed for operational services such as earth observation and internet services with a constellation of satellites. In these cases, the satellite's mission life may be greater than two years and the product assurance (PA) plans need to be drawn on par with bigger operational satellites. Hence, the overall PA methodology followed for an operational satellite is described below and PA plans to be adopted for a Nanosatellite project by educational institutions can be a subset of this overall plan.

11.8 PRODUCT ASSURANCE ELEMENTS

Product assurance entity encompasses almost all elements of a Nanosatellite project including:
- Electrical, electronic and electromechanical components (EEE)
- Materials and processes
- Quality control aspects
- Testing and evaluation
- Reliability estimation
- Quality of software
- Mission aspects
- Project reviews
- Non-conformance management

Always keeping in mind the goals of the mission, the role of the PA team during various stages can be summarised as:
- Satellite configuration definition
- Sub-system design and analysis
- Procurement of EEE components and materials
- Sub-system hardware fabrication/development
- Test and evaluation of each sub-system
- Satellite level assembly, integration and testing activities
- Satellite level environmental tests

The PA team needs to be closely associated with each of the above activities and ensures strict adherence to quality guidelines. Specific activities of PA in the above phases are briefly described below.

11.8.1 Procurement of EEE component and materials

Electrical, electronic and electromechanical (EEE) components are the 'building blocks' of a satellite. These include active components (transistors, diodes, VLSI devices), passive components (resistors, capacitors, inductors, magnetic cores) and electro-mechanical components (relays, switches). They are available in different quality grades – space-grade, military-grade, industrial grade and commercial-off-the-shelf (COTS). Selection of these components for a Nanosatellite involves a trade-off among factors such as reliability, available budget and schedule. Most of the Nanosatellites developed by educational institutions use COTS components due to their low cost and ready availability whereas operational Nanosatellites may use space-grade or military-grade components. Procurement of these components is an important activity in the satellite project and the PA team plays a vital role by preparing a 'preferred components list' (PCL) available from reputed vendors, providing procurement specifications, reviewing the heritage, test data and conducting additional screening tests. More details on the procurement strategy of COTS components for a Nanosatellite are given in **Annexure-A14.**

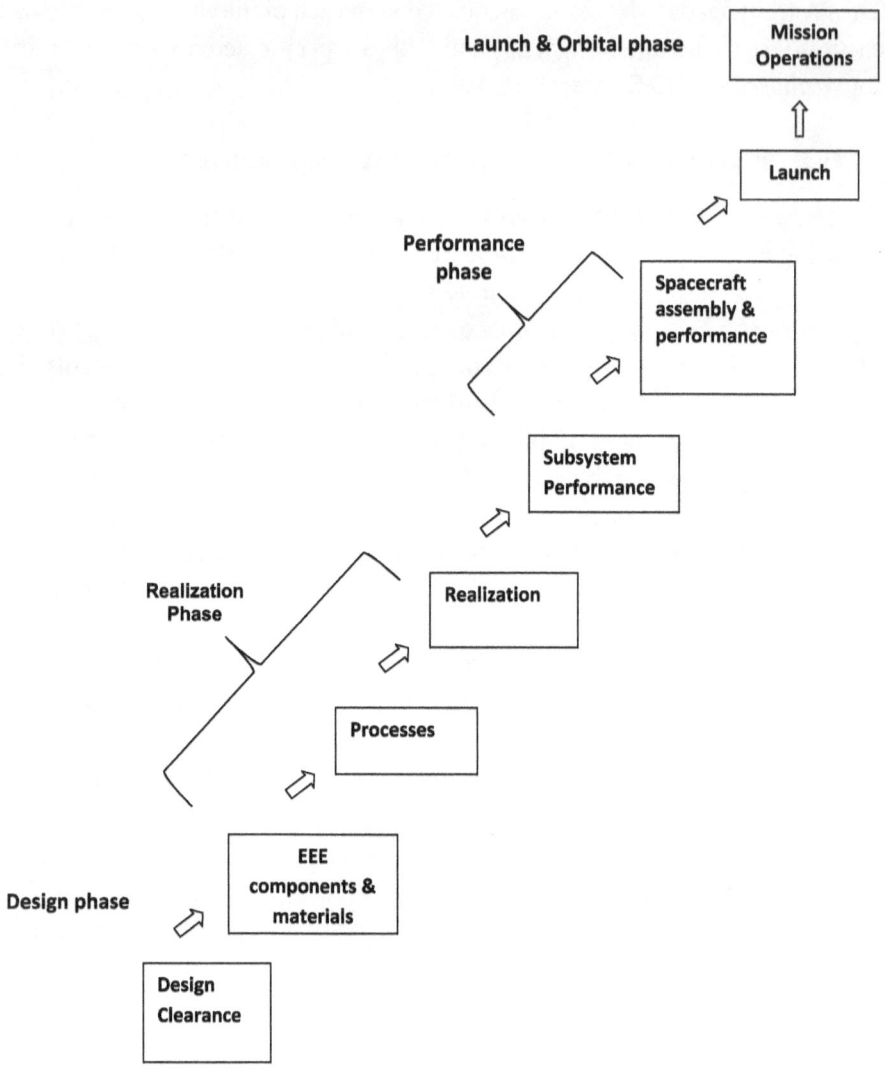

▲ **Fig. 11.3:** Product assurance phases

11.8.2   Sub-system fabrication stage

Fabrication of satellite sub-systems is the most critical phase and strict surveillance by the PA team is mandatory to avoid any surprises or failures during later stages. Prior to the start of fabrication, the PA team has to ensure that all the processes are qualified, personnel are trained, external vendors are certified and all process documentation is available. In-process control with online inspections, various checkpoints and sample tests need to be identified as well as implemented. It is also important to conduct periodical audit processes of fabrication facilities, vendor premises and inspection reports by the PA team. Any deviations observed during fabrication shall be immediately reported to the concerned agency for review and disposal.

11.8.3   Test and evaluation stage

After fabrication, all sub-systems have to undergo test and evaluation (T&E) to verify their performance against their specifications and the PA team will be the final authority to certify the sub-system as fit for integration with the satellite. Major steps involved in the T&E phase to be addressed by the PA team are:

- Preparation of T&E plan and review by experts
- Finalisation of test specifications, including environmental tests
- Calibration and certification of all test equipment
- Certification of environmental test facilities' readiness
- Verification and validation (V&V) of onboard software modules
- Compilation of test results and log books
- Record of non-conformance reports, reviews and disposals with waivers
- Issue of T&E certificates with special precautions, if any

A typical sequence of sub-system level T&E phase is shown in **Fig. 11.4.** In addition to these tests, it is important to conduct a 'hardware-in-loop simulation' (HILS) test on ADCS elements on a three axes servo table to verify the correctness of control algorithms, error polarities and electrical interfaces. With the final T&E certificate, each sub-system will be delivered for satellite integration.

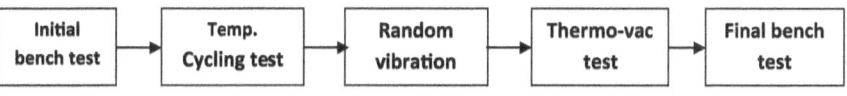

▲ **Fig. 11.4:** Sub-system T&E sequence

11.8.4 Satellite assembly, integration & test stage

A satellite's assembly, integration and test (AIT) get initiated well before the sub-systems are delivered to ensure the smooth progress of the AIT phase. These activities include, i) keeping all mechanical fixtures, handling systems and tools ready; ii) fabricating the electrical wire harness, testing it and installing it on the satellite structure; iii) keeping all test equipment ready after calibration; iv) keeping test plans and testing procedures ready; and v) ensuring that all checkout software modules are installed on computers after V&V.

The sub-systems will be integrated into the satellite in a pre-defined sequence and the fully assembled satellite will undergo environmental tests. At the end of successful completion of this sequence of tests, the satellite will be ready for shipment to the launch base. During this phase, the PA team will be actively involved in 'monitoring' the progress as per laid down guidelines and specifications. Without getting influenced by the 'schedule pressure', the PA personnel shall independently ensure the safety of the satellite and persons working on it, cleanliness and contamination control and all the planned activities/tests are completed without any short cuts.

11.8.5 Satellite level environmental tests

A fully assembled satellite goes through AIT that includes environmental tests to verify the satellite's condition to withstand and perform its functions after it is subjected to simulated launch and space environment. These tests include sine vibration, random vibration, thermal-vacuum and acoustic test (Nanosatellites are not put through the acoustic test). Sine and random vibration tests simulate launch loads at lift-off, trans-sonic accelerations and launch vehicle stage separations. Vibration tests are carried out on electro-dynamic shakers of requisite capacity on all the three axes of the satellite. The dynamic response of the satellite systems during vibration tests is monitored using a set of accelerometers with one or two of them in the control feedback loop of the shaker. Typical specifications for these vibration tests are shown in **Table. 11.1** and a Nanosatellite under vibration test is shown in **Fig. 11.5**.

▼ **Table. 11.1:** Typical vibration specifications

Sine vibration test

	Frequency range (Hz)	Qualification level	Acceptance level
Longitudinal axis	4 – 10	10 mm (0-peak)	8 mm (0-peak)
	10 – 100	3.75g	3g
Lateral axis	2 – 8	10 mm (0-peak)	8 mm (0-peak)
	8 – 100	2.5g	2g
Sweep rate		2 Octave/minute	4 Octave/minute

Random vibration

Frequency (Hz)	Qualification PSD (g²/Hz)	Acceptance PSD (g²/Hz)
20	0.002	0.001
110	0.002	0.001
250	0.034	0.015
1000	0.034	0.015
2000	0.009	0.004
gRMS	6.7	4.47
Duration	2 minutes/axis	1 minute/axis

▲ **Fig. 11.5:** Nanosatellite under vibration test (Courtesy: https://makesat.com)

▲ **Fig. 11.6:** Nanosatellite under thermo-vacuum test (Courtesy: ISRO)

Thermo-vacuum test of a satellite is carried out in a special, suitably sized chamber. Usually, Nanosatellites are tested in 2 to 3 m dia chambers as shown in **Fig. 11.6**. These chambers can simulate vacuum of the order of 10^{-6} torr or mbar and temperature extremes of about –180 degree Centigrade to + 100 degrees Centigrade through the circulation of liquid nitrogen and gaseous nitrogen. The satellite under test is placed either in a hanging position inside the chamber using insulated steel ropes or placed on a thermally controlled table inside the chamber. A typical sequence of temperature cycles operated during the thermo-vacuum test is shown in **Fig. 11.7**. It should be ensured that during the cycles and soaks, all satellite sub-system temperatures reach their specified upper and lower limits. Performance of satellite systems is verified during hot and cold soaks as well as transitions between cycles.

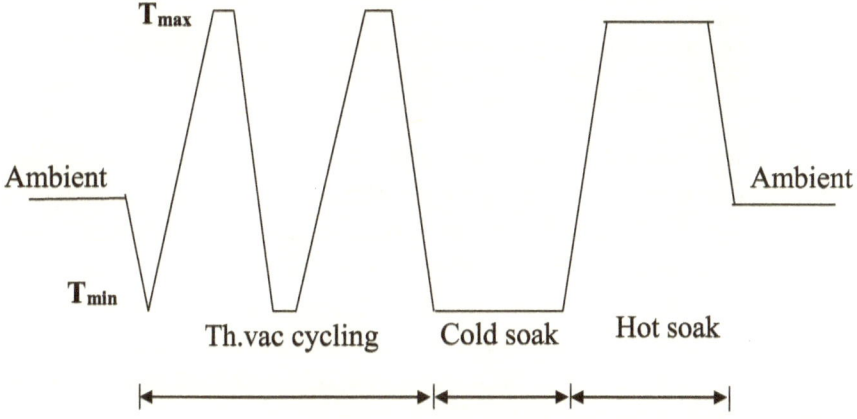

▲ **Fig. 11.7:** Typical thermo-vacuum cycles for Nanosatellite

After successful completion of environmental tests, a Nanosatellite undergoes the final functional checks and is containerised for transportation to the launch base. During the environmental tests, the PA team ensures that the satellite is subjected to specified levels of vibration and temperatures, its performance meets the specifications, the satellite is handled safely and all precautions are taken for its safe transportation. It is also important that the PA team documents all the test results, deviations, their disposal as well as critical observations and presents them to the satellite flight readiness review or pre-shipment review committee for clearance. It is imperative to ensure a strong product assurance strategy in a satellite project to develop a reliable satellite.

11.9 SOFTWARE QUALITY ASSURANCE

With increasing reliance on embedded systems in Nanosatellites, the role and complexity of onboard software are also increasing, needing more stringent quality assurance strategies. Even though the process for software assurance generally remains the same across different classes of satellites, the rigour of its adaptation could vary depending on mission objectives.

Software assurance aims to verify and affirm bug-free functionality for various aspects such as spacecraft dynamics, closed-loop functions, mission operational sequences, timing margins, data bus transmissions, input-output protocols and onboard autonomy.

> Dilution of system requirements to lower tier tasks - design, architecture and ultimately to software code is a sequential 'water fall' process, to be astutely followed by software developers Properly 'verified' code does exactly what it was designed to do But, what the code does may not be the right thing, and hence 'validation' is required

Most common methods of software assurance involve simulation of data of various functional elements such as attitude sensors, ADCS actuators, power safe logics, telemetry/ command, fault isolation and detection. Good verification demands good fidelity of simulation and evidently comprehensive simulations are highly resource-intensive. However, for a Nanosatellite, the rigour of testing cannot be so exhaustive and the fidelity for test simulations would be at reduced effort level. It is also an acknowledged practice to use multiple desktop computers and public domain tools to deal with simulations and asynchronous inputs.

Continuous testing being the key for a Nanosatellite, the software is required to accommodate specific features for automated tests. Thus, open-source software holds a lot of promise for software developers. Public domain scripts like MATLAB™ and several C-macros are extensively used, which can seamlessly handle computational requirements as well as bi-directional interactions for the processor devices.

Some of the key factors to obtain reliable software include:
- Preference to use an open-source operating system where requirements can be compiled and run correctly on a desktop system. The underlying aspect here is that with a well verified operating system exhaustive validation may not be essential.

- Notwithstanding the validation and testing of the software on the ground, the provision for flexibility in uploading software (boot-loading codes and computational applications) builds in a Nanosatellite helps to re-programme and re-configure processors and interfaces. This feature forms a critical test case during software testing.
- Fault tolerance features shall not be over-catered, and instead, they need to be logically sufficient. Increased branches of fault-tolerant paths result in unwarranted wastage of expensive computing resources.
- Ensuring reliable communication among processors, memory chips and other participating interfaces is a prime factor for the software.

11.9.1 Onboard software

The software that is a part of the satellite and flies with the hardware components is called onboard software. The onboard software generally handles attitude determination and control (ADC), navigation and guidance (NG), telecommand and processing (TCP) and telemetry (TM). Generally, the onboard software is written in high-level languages namely Ada or C. If the C language is used, care needs to be taken throughout the software development life cycle (SDLC).

Nanosatellites generally do not have high computational power in onboard systems due to limitations of resources. Also, the processor devices are of COTS category and possibilities of data errors due to radiation-induced effects are very high. Hence, software of a Nanosatellite, besides the required functionality, must also address fault handling, mitigation and alternate operational modes.

11.9.2 Ground software

The software that is used on the ground to design, test and validate sub-systems of the satellite is called ground software. Ground software also includes telemetry data processing and display, command up-linking and payload data processing during mission operations. Any programming language can be used for the ground software.

11.9.3 Quality Assurance (QA) for software

Quality Assurance for software is applicable for both the onboard and ground software. QA is more imperative for the onboard software since any bug in the software may lead to mission failure. Different techniques and methods followed in software assurance are briefly described below:

11.9.4 SDLC models for a Nanosatellite

Following models of software development life cycles (SDLC) have emerged in space applications and specifically for a Nanosatellite:
• Modified waterfall model
• The agile method with test-driven development (TDD)
• VALidation and VErification for Embedded Systems (VALVES) using Tools and Techniques Set (TTS)
• Model-based software design and verification

11.9.4.1 Modified waterfall model

A modified waterfall iterative model is best suited for space applications because, in the majority of the cases, the software requirement does not freeze; it evolves and changes. A typical modified waterfall model is shown **in Fig. 11.8.** Compared to a pure waterfall model, a modified waterfall model has advantages such as better flexibility, less documentation and implementation of easy areas need not wait for hard areas. Disadvantages of this model are that the milestones are more ambiguous and parallel activities may cause miscommunication as well as misinterpretation.

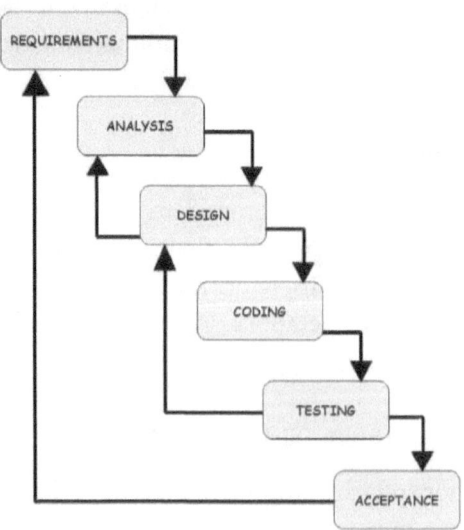

▲ Fig. 11.8: Modified waterfall model

11.9.4.2   Agile method with test-driven development (TDD)

The agile method with test-driven development (TDD) is found to be very effective as this approach blends 'testing' and 'software development' together. The main features of this method are:
- Focus on the code rather than the design
- Based on an iterative approach to software development
- Intended to deliver working software quickly to meet changing requirements
- Reduced overheads in the software process (e.g. by limiting documentation)

The principles of agile methods are:
- *Customer involvement*: Customers are closely involved throughout the development process. Their role is to provide and prioritise new system requirements and evaluate the iterations of the system.
- *Incremental delivery*: Software is developed in increments with the customer specifying the requirements to be included in each increment.
- *People, not process*: The skills of the development team are recognised and exploited. Team members are left to develop their own ways of working without prescriptive processes.
- *Embrace change*: Expect the system requirements to change and so design the system to accommodate these changes.
- *Maintain simplicity*: Focus on simplicity in both the software being developed and in the development process. Wherever possible, actively work to eliminate complexity from the system.

Test-driven development (TDD) is a software development process that relies on the repetition of a very short development cycle. Requirements are turned into very specific test cases and then the software is improved to pass new tests.

11.9.4.3   VALVES using TTS

Flexible test environment helps 'fast feedback' to the changed or added features for achieving a faster time-to-market. Using tools and techniques set (TTS) for developing complex embedded systems called VALVES is found to be highly relevant. VALVES is essentially an evolution of the basic TTS.

11.9.4.4 Model-based software design and verification

Small and Nanosatellite literature reveals that an emerging trend is to use automation to create detailed software designs from high-level graphical inputs and then use automatic code generation to create the code. This process is called model-based design (MBD) or model design-driven (MDD) software development.

Model-based design tools offer no help with getting the software requirements right but they do help alleviate much of the burden to make sure the software meets the requirements. The model-based design approach begins with a careful delineation of the software requirements, just as is done for the waterfall development process. However, after the textual requirements are complete, the MBD approach creates the high-level software architecture in the form of graphical models of the code that shows connections between program modules and describes the software behaviour using state charts, sequence diagrams, and/or timing charts. Model-based design approach requires the top-level software architecture to encapsulate the software function within individual modules. This requirement is imposed to guarantee that each module can function without dependence on other modules. The modules interface with each other so that control signals and data are passed between the modules. This graphical modelling of the software at a high level is usually better comprehensible to other design engineers trying to understand the function of the code and hence promotes sharing as well as code re-use. It's much easier for a new user to comprehend the function of the code from a block diagram than reviewing the commented code.

Once the software architecture is specified in the model notation, the model-based design tools perform the detailed software design without the user having to worry about declaring variables, allocating memory or anything else normally associated with writing a Java, C or C++ program. The detailed software design created by the model-based design tools may be the actual code but it can also be a meta-model that can be executed to verify that the software design meets the software requirements before the actual code is generated. This allows verification that the expected high-level behaviours of the software are actually triggered by high- level mechanisms. It can be executed to verify that all states of the system are reachable and that all communication interfaces work properly.

Model-based design tools use the process of automatic code generation to keep the model in sync with the code. The code is automatically generated from the detailed software design or meta-model. Automatic code generation thereby ensures coherence between the model and the code, which is important for software engineers desiring to re-use model elements in new programmes.

There are many model-based design tools available. However, the model-based design tools that most stand out in the small and Nanosatellite literature are the Unified Modelling Language and the MathWorks.

11.9.5 Software coding

Software design is followed by the process of coding. This coding demands adherence to the standard coding guidelines. They are:

- C style guide (from Software Engineering Laboratory Series SEL-94-003) – National Aeronautics and Space Administration (NASA), Goddard Space Flight Centre, Greenbelt, Maryland 20771
- MISRA – C coding guidelines 2016 (MISRA: Motor Industry Software Reliability Association)

The programmers involved in coding shall follow naming conventions for variable names, modules, sub-modules and files. Compliance to MISRA-C shall be confirmed using the tool POLYSPACE or any equivalent tool.

11.9.6 Software testing, verification and validation

Software testing is highly crucial to produce *bug-free* software. Types of software testing and various disciplines of testing are listed below:

- White-box testing
- Static testing
- Static testing by humans
- Static testing using tools
- Structural testing
- Unit testing
- Code coverage testing
- Code complexity testing
- Black box testing
- Positive testing
- Negative testing
- Boundary value testing
- Integration testing
- Regression testing

White-box testing

This is also called a *clear box*, *glass box* or *open box* testing. White box testing is testing by looking at the programme code. This testing takes into account the program code, code structure and internal design flow. This testing will help to reduce the delay between the injection of the defect and its detection. Since the programme code represents what the product does rather than what is intended, white box testing makes us understand what the product is actually doing.

Static testing

Static testing requires only the source code neither the binaries nor executables. This does not involve executing (running) the programme on a computer (processor). On the other hand, it involves examining the code to find out whether the code:

- Works according to the functional requirement
- Has been written as per heritage
- Missed out any functionality
- Handles the errors properly

Static testing by humans

This method relies on the principle of software experts reading the programme code, instead of executing it, to detect errors. This method has its own advantages. For example, if there are two similar variable names, using one in place of the other may go undetected by the code execution method, whereas this gets detected in the manual reading mode. Code walk-through is one such method of static testing by humans.

Static testing using tools

Standard tools are used for static testing. Again it is the source code subjected to the testing (not its binary equivalent or executable code). POLYSPACE is one such tool that can be used for static testing or any equivalent tool can be used.

Structural testing

In contrast to static testing, structural testing is done by executing the code on a host computer and a target computer. The programmer or tester will, *a priori*, design test vectors, which are given as inputs and the programme is executed to see the computed results. It is mandatory that the computed values match the expected results in the case of Boolean values and integers, whereas in the case of real values, they remain within tolerance limits.

Unit testing
Unit testing is one way of structural testing. This is conducted for one programme unit or one module at a time.

Code coverage testing
Code coverage testing is one attribute of structural testing. This means the test vectors should be designed such that after completing a set of test cases, the code coverage is 100 per cent. However, there are tools helpful to notify the percentage of code coverage achieved.

Path coverage testing or code complexity testing
It is important to ensure that path coverage during the tests is 100 per cent. The minimum number of test cases required for 100 per cent path coverage is dictated by the cyclomatic complexity.

Black Box testing
Black box testing is done based on requirements and encompasses end-user perspectives. It addresses stated requirements as well as implied requirements.

Positive testing
Positive testing aims to make sure that the software processes are as expected. Testing the software to deliver an error message when the error message is expected is also positive testing.

Negative testing
Negative testing aims to check that the software does not fail when an unexpected input is given. The purpose of this is to try and break the system. This covers scenarios for which the software is not designed and coded. In other words, the input values may not have been represented in specifications. These test conditions may be termed as unknown conditions for the specifications.

Boundary value testing

In this case, the test conditions will be designed around the boundary values of the valid range.

Integration Testing
Software integration testing

Contrary to the unit testing, in this case, tests are to be conducted with logically related two or more software units at a time to ensure that the arguments are properly passed amongst them.

Hardware-software integration testing

All the tests mentioned so far can be done on a host computer with the same version of the compiler whereas the hardware-software tests will be conducted with hardware platform and the flight processor.

Regression Testing

Changes in the software are inevitable due to any of the following:
- Change in requirement
- Enhancement in requirement
- Correct implementation of the functionality
- Fixing the discovered bugs

Whenever a tested software changes, it is important to ensure that newly added/changed piece of software has not affected the functionality of the tested software. Tests conducted to ensure this non-degradation is called regression testing. Depending on whether the software has been added or removed or changed, the test case (which is a sub-set of the earlier test cases) is to be determined depending on where the code is disturbed.

Verification and validation (V&V)

Verification takes care of activities to answer the question, *"Are we building the product right?"* and validation takes care of activities to answer the question *"Are we building the right product?"*

Verification

To build the product right, certain procedures are imposed at the beginning of the life cycle. The purpose is to prevent defects. Reviews held at various stages of SDLC are examples of verification activities. Verification is termed as quality assurance QA.

Validation

Once the product is built, certain activities are carried out to validate the product's specifications. Tests carried out are examples of validation activities. Validation is termed as quality control (QC).

11.9.7 Software configuration management (SCM)

Software configuration management involves the following:

- To identify the elements of the project that fall under configuration management
- To control the changes in software by addressing its requirements, design or code and review to ensure that new change will not affect the performance
- As and when changes occur in the source code, it is important to maintain the **version number.** Good tools are available in the market to maintain version numbers
- Auditing of above mentioned three activities
- Working out a mechanism to be put in action to appraise changes to all concerned teams

11.9.8 Maintenance/code patching/remote programming

The term "maintenance" is relevant only for ground software. In the case of ground software, the bugs detected are removed in the maintenance phase. However in the case of onboard software also bugs detected may be wiped out by code patching or remote programming. That is a unit of software or a module that is defective is identified and the code of the software with the correction is used to simulate the satellite on the ground to confirm that the changes are proper. The executable code of the corrected software is uploaded to the satellite as telecommands. The address of the calling routine (in the called routine) is suitably modified to call the corrected module instead of a defective module. This process is called remote programming.

REFERENCES

1. 'Software QA in Space Applications (PPT file) – as part of SQA Course for Group A officers at DIQA, Bengaluru' by B.V. Prasad, S. Alagu Rani and K.S. Prasada Kumari
2. 'Formal Methods and its Applications for V&V of Onboard Software', 3rd International Conference on Reliability and Safety Engineering December 17-19, 2007, Organized at Udayapur by IIT Kharagpur.

3. *'Software Engineering – A Practitioner's Approach'* by Roger S. Pressman, Published: McGraw Hill International Edition
4. *'Software Testing Principles and Practices'* by Srinivasan Desikan and Gopalaswamy Ramesh
5. Software Testing https://www.Guru99.com
6. *'Creating Capable Nanosatellites for Critical Space Missions'* by Aaron Q. Rogers and others
7. *'Developing and Testing Software for the 14-BISat Nanosatellite'* by Rogerio Atem de Carvalho and Milena et al
8. Chapter3 - *'Agile Software Development Lecture1'*
9. *'Survey of Verification and Validation Techniques for Small Software Development'* by Stephen A. Jacklin

* * *

L.S. Satyamurthy has worked as a satellite quality system expert since the beginning the Indian satellite program. He has extensive experience in international technical liaison, space technology applications in health care and commercialisation of space products and services. He has served as Scientific Diplomat at the Embassy of India in USA, Director, Business Development at Antrix Corporation and as Program Director, Telemedicine, ISRO.

B.V. Prasad is an applied mathematician and software quality system expert, with vast experience in the areas of computer software for supporting control system group for simulation in the design and development of attitude control systems of remote sensing and communication Satellites. He has also been responsible for quality assurance of satellite on-board software.

ASSEMBLY, TESTING & INTEGRATION

M. Venkata Rao & A.A. Bokil

Assembly, integration and test (AIT) is a multi-disciplinary activity that requires expertise from different engineering domains and calls for close interaction between the AIT team, satellite system designers and launch vehicle group. Generally, there are certain pre-requisites for the smooth conduct of AIT activities with respect to the configuration of a Nanosatellite's electrical and mechanical interfaces, which shall be defined properly and documented. Primarily, it shall include the satellite interface with the launch vehicle separation mechanism, which needs to be prepared meticulously, reviewed and implemented.

AIT activities broadly encompass three areas, mechanical integration, electrical integration and ground checkout of the complete satellite and environmental tests covering thermo-vacuum, vibration, communication and antenna tests as applicable for Nanosatellites. Electromagnetic interference and electromagnetic compatibility are two important parameters to be taken care of in the Nanosatellite design to minimise the interference effects on various sub-systems.

Implementation of proper grounding scheme during electrical integration is essential to reduce noise interference. Unlike bigger satellites, Nanosatellites being small are less susceptible to ground loop currents and can adopt simplified grounding techniques.

The above aspects are covered in detail in this chapter for the benefit of the reader.

Fabrication, test and evaluation of sub-systems such as power, TTC, ADCS, OBC and payload for Nanosatellites will be followed in the next phase of activity commonly referred to as assembly, integration and testing (AIT). The AIT phase, which extends up to launch phase, involves a series of assembly activities and sequence of tests to be carefully planned and executed on the satellite. AIT is a multi-disciplinary activity that requires expertise from different engineering domains and calls for close interaction among the AIT team, sub-system designers and the launch vehicle team.

There are certain pre-requisites for the smooth conduct of AIT activities as given below:

- The configuration of a Nanosatellite including all electrical and mechanical interfaces is well defined and documented
- Test plans and procedures, test cases and input simulations are finalised after going through reviews by domain experts
- All test equipment, power supplies and simulators are ready for use after going through their T&E and calibration
- All required software modules for AIT such as telemetry and telecommand database, data acquisition, processing and archival are installed on computers after going through necessary verification and validation tests
- Required mechanical equipment and hardware, including tools, handling systems and transportation container are available
- Mechanical assembly drawings for all interfaces duly approved by all concerned are available
- Launch vehicle interfaces, including Nanosatellite ejection mechanism, are defined and documented

AIT activities broadly encompass three areas: Mechanical integration, electrical integration and ground check, which are closely interleaved in sequence. Major tasks/responsibilities of these three areas are briefly covered below.

12.1 MECHANICAL INTEGRATION (MI)

Mechanical integration activities start right from the definition of the Nanosatellite configuration in close interaction with the structure and thermal design teams. A typical set of MI activities are as given below:

1. Preparing the layout of mounting locations for all satellite sub-systems and appendages such as antennae and payload on the satellite structure is the first task in mechanical integration. While preparing these layouts, it is essential to address the functional requirements of sub-systems along with electrical, thermal, structural and operational requirements. These requirements include mechanical clearances for accessibility and ease of assembly/disassembly operations, proper surface contact at the base for good thermal contact, cut-outs for routing of harness wires/cables, unobstructed field-of-view for optical sensors and RF antennae. Once the layout is finalised, mechanical drawings for the layouts are prepared and reviewed by all concerned. Using CAD software packages such as

Autocad, a 3-D model of the Nanosatellite will be generated to visualise the assembly clearances. A typical 3D CAD model of a Nanosatellite with sub-system locations is shown in **Fig. 12.1**.

2. Design and fabrication of mechanical housings and brackets required for sub-systems assembly and procurement of assembly tools and hardware in advance

3. Interaction with launch vehicle team to finalise the Nanosatellite's mounting interface on the launch vehicle and its deployment mechanism, launch base operations and schedule

4. Estimation and measurement of physical parameters of the Nanosatellite, namely its mass, the centre of gravity and moment of inertia around all three axes –need to be provided to the launch vehicle team

5. Design, fabrication and qualification of various handling systems, jigs and fixtures required for AIT phase activities, including a suitable container for transportation of the Nanosatellite to test facilities (vibration test and thermo-vacuum test) and to the launchpad

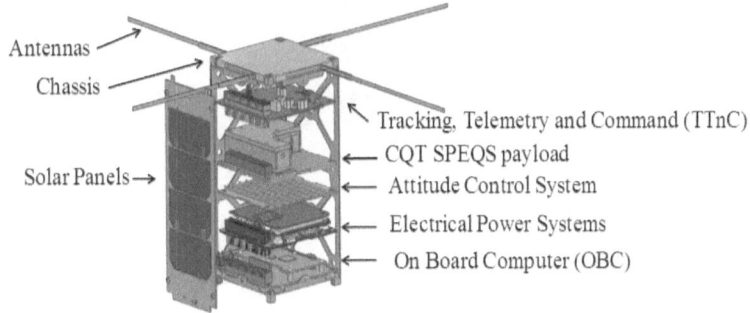

▲ **Fig. 12.1:** 3D CAD model of a Nanosatellite

12.2 ELECTRICAL INTEGRATION (EI)

Electrical integration activities also begin with the finalisation of the Nanosatellite's configuration and interface definition of all sub-systems. Major activities of electrical integration are:

• Sub-system layout preparation
• Wire harness design, fabrication and qualification
• Sub-system level/system-level integration and testing
• EMI/EMC/ESD analysis and testing
• Implementation of proper grounding scheme

- Planning and execution of environmental such as vibration and thermo-vacuum tests
- Pre-launch activities at the launch base

A brief summary of these electrical integration activities is given below:

1. Preparation of a layout for sub-system placement on the Nanosatellite in close interaction with the mechanical integration team will be followed by harness fabrication. During this exercise, the EI team will carefully address issues such as the type of signal or power lines and their possible interference with adjacent lines, type of cables and connectors to be used, shielding of RF lines to minimise radiation leakages, routing of harness for ease of accessibility. Different types of cables normally used in a Nanosatellite are shown in **Table. 12.1.** The harness is fabricated based on the electrical interface details provided by sub-system designers in the EID form (electrical interface data) document and it is thoroughly tested for pin-to-pin compatibility as well as electrical continuity.

▼ **Table. 12.1:** Types of cables used for harness

SI No.	Signals	Cable type
1	DC raw power	Twisted pair
2	Low-frequency digital/ analogue	Twisted pair (TP)
3	Low level analogue	Shielded twisted pair (STP)
4	High-frequency digital	Coaxial
5	RF signal	Coaxial
6	Electro-explosive device signals	Double shielded twisted pair

2. Sub-systems are assembled on the Nanosatellite structure and tested for their performance when integrated with other onboard systems. Test procedures and test equipment required are to be evolved, reviewed and kept ready *a priori* for this activity. Different sub-systems get integrated sequentially to form a compatible integrated system. Once the performance of all sub-systems and payload is verified in the dis-assembled mode, the Nanosatellite is completely assembled, including various appendages such as RF antennae, solar panels and thermal control elements and finally tested in the assembled mode.

3. Electromagnetic interference (EMI) and electromagnetic compatibility (EMC) are two important parameters to be taken care of in the Nanosatellite to minimise the noise effects on sub-systems. While the EMI/EMC effects

can be theoretically analysed using suitable software, several practical solutions need to be implemented during electrical integration such as:

- Power, signal and RF cables are to be bundled separately and routed away from each other
- Inductive and capacitive coupling between adjacent wires can be minimised by using twisted pair cable and shielded wires respectively
- Power and signal wires preferably should not share the same connector or they should be well separated in the pin allocation in the connector
- In case of shielded cables, the shield should be connected to chassis of the connector
- Implementation of a proper grounding scheme to avoid ground loops and current mismatch (see **Annexure A-13** for details on the grounding scheme)

Apart from taking above precautions, sub-system level EMI/EMC tests are to be conducted as per MIL-Std-461 to establish margins with respect to expected levels on the satellite platform.

Electro-static discharge (ESD) is a hazardous phenomenon, which can cause the failure of electronic devices if touched with bare hands. To avoid these failures, the following precautions are to be taken during AIT:

- Wearing anti-static wrist straps firmly grounded while touching electronic device pins with probes
- Cleanroom/laboratory where the Nanosatellite's AIT activities take place, the room should be covered with anti-static flooring
- Sensitive items such as wired PCBs and electronic packages should be handled always in anti-static bags

4. Implementation of proper grounding scheme during electrical integration is very important to avoid component failures and reduce noise pickups. Unlike bigger satellites, distances between sub-systems are small in a Nanosatellite thereby reducing the ground loop currents and simplifying the grounding techniques. Following guidelines are suggested for effective grounding:

- In case of a distributed power system, the bus return wires to be connected to the structure
- All RF cables to have multiple groundings
- Shields of the shielded twisted pair wires are to be connected to the connector shell with minimum pigtail length
- Proper electrical contact between different sides (panels) of the Nanosatellite structure to be ensured using thick copper braids

5. An assembled Nanosatellite will undergo environmental tests, namely, vibration and thermo-vacuum (TVC) tests. Since the Nanosatellite will be in power OFF condition during the launch, it is not powered ON during the vibration test and its performance is evaluated before and after the vibration test. However, in case of the thermo-vacuum test, the Nanosatellite is powered ON during testing and hence requires harness wires of sufficient length between the TVC chamber and the test lab. These cables need to be fabricated in consultation with test facilities team and quality assurance team.

6. After the transportation of the Nanosatellite to the launch base, it needs to be powered ON and tested as a final check before launch. The same test equipment and harness cables used for assembled mode testing will be taken to the launch base. In the case of launch by the PSLV of ISRO, a Nanosatellite is kept in power OFF condition after mounting on the launch vehicle and it will be ON only after injection into the orbit. The typical mounting configuration of Nanosatellites on the PSLV launch vehicle is shown in **Fig. 12.2**.

12.3 GROUND CHECKOUT (GCO)

The ground checkout team is responsible for carrying out the integrated satellite testing (IST), including the acquisition of satellite telemetry (housekeeping parameters) and payload data and analyse the same to verify the performance of all sub-systems. The GCO team, in close interaction with mechanical and electrical integration teams, plans the sequence of tests – disassembled mode IST, assembled mode IST, environmental tests, deployment tests, solar array illumination checks and RF radiation checks. A typical sequence of AIT activities is shown in **Fig. 12.3**. Major activities of the GCO team are summarised below:

1. Identify the test equipment required for various tests, design and fabricate/procure them, carry out T&E and ready them well before the start of IST. These equipment include:
   - Solar array and battery simulators
   - Data acquisition systems for telemetry and payload data
   - RF transmitters, receivers and demodulators
   - Telecommand encoder
   - Stimuli generator for sensors and ADCS
   - Data archival and storage system
   - Computer systems for testing operations

- Laptop with relevant interfaces and necessary software modules programmed to carry out most of the above

2. Prepare procedures for each type of test and generate the step-by-step sequence of commands to be uplinked and telemetry data to be verified. For this purpose, a database containing a list of commands with codes and a list of telemetry channel details should be prepared and installed in the test computer. This database should be updated 'as and when' there is any change in the TM/TC lists. These test procedures need to be reviewed and vetted by sub-system designers and product assurance experts.

3. All the software modules required for carrying out operations: acquire, process and verify TM and payload data to be developed, after going through verification and validation tests and implemented on the test computers.

4. Conduct the IST on the Nanosatellite in the disassembled and assembled modes, thermo-vacuum test and finally at the launch base, check the performance and give the 'go-ahead' for the launch.

The AIT is basically a set of multi-disciplinary activities and operations. Though these activities are described separately for electrical integration, mechanical integration and ground check-out teams, a small team with system engineering approach and expertise can carry out AIT activities for a Nanosatellite.

With a focussed effort of the AIT team and sub-system designers along with good planning and execution, any Nanosatellite mission can be successfully accomplished.

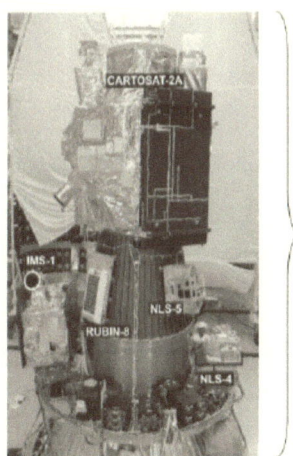
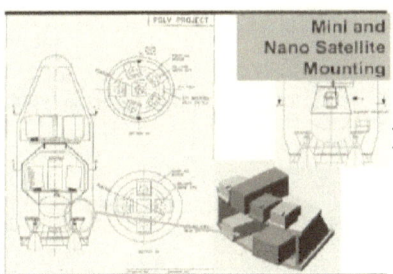

▲ **Fig. 12.3:** Mounting of Nanosatellites on a launch vehicle (Courtesy: ISRO)

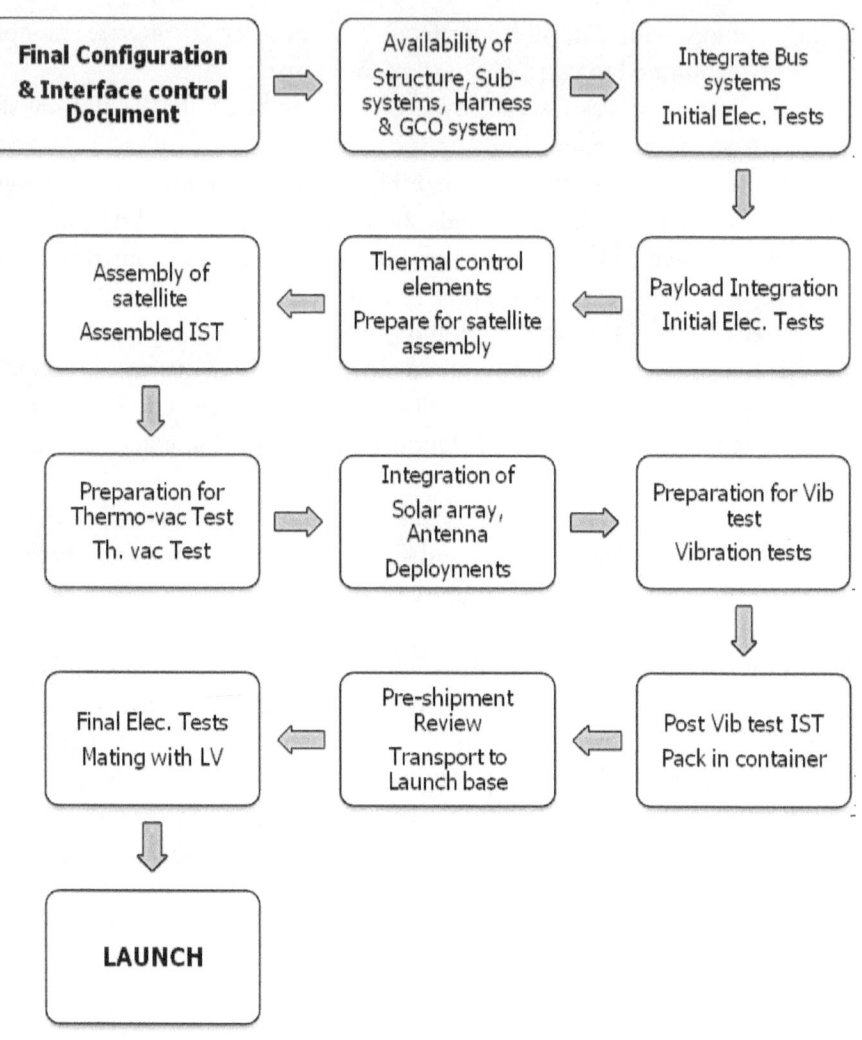

▲ **Fig. 12.4:** A typical sequence of AIT

REFERENCES

1. *'Development Plan of Harness Model for Small Satellites'* by Muhammad Fahad Hassan, Muhammad Atif et al, International Conference on Aerospace Science & Engineering (ICASE)-2011
2. NASA-HDBK-4001

* * *

M. Venkata Rao has more than 30 years of experience in system engineering aspects of electro-optical systems and earth observation payloads and project management of Indian remote sensing satellites program. He has served as Project Director for two of ISRO's Remote sensing satellites.

A.A. Bokil has three decades of experience in the field of assembly, integration and testing (AIT) of satellites in his career at ISRO. He was trained in ESA/ NASA during SPACELAB program. He designed and tutored an academic course on satellite architecture for the ME programme at Indian Institute of Science, Bengaluru. He has led a team of engineers for a technology exchange programme between India and Iran. He also worked as Director Space Segment, Broadcast Operations and Telemedicine Operations.

CONCLUSION

Satellite technology is a multi-disciplinary field covering science, technology, engineering and mathematics (STEM), which is challenging to understand, analyse and design in the holistic space system. The Nanosatellite is a frontier technology marvel, which can create interest in students of schools, colleges and universities/institutions. Further, this technology is at the easy reach of start-up companies/industries, whether it is for building a satellite or ground systems or the various application services to the end-user/customer.

After a very assiduous and focused browsing of the previous 12 chapters of this book on satellite technology in general and Nanosatellite in particular, we hope that the readers are exposed to the nuances of satellite technology covering the scientific and engineering aspects of system design, manufacturing and test, preparing a satellite for launch and subsequent service.

Some of the universities/institutions in India and abroad have built and launched Nanosatellites for remote sensing, communication and scientific applications. They have derived knowledge from various books like the one which is presented to the readers here. If this work of ours has instilled some understanding and confidence in the readers, we would consider our work truly successful and worth the effort.

ANNEXURE - A1
ORBITAL MECHANICS

To describe the motion of a satellite, we assume that it is being acted upon only by the gravitational pull of the spherical earth and ignore all other objects in the universe. These have only a secondary influence on the satellite's motion and hence can be treated as perturbations to the solution obtained for the two-body problem.

Newton's second law of motion states, $\mathbf{F} = \mathbf{ma}$ where $\mathbf{m}$ is mass of the satellite and $\mathbf{a}$ its instantaneous acceleration.

The earth's gravitational force on the satellite of mass $\mathbf{m}$ is given by

$$- (GMm/r^2) \, \hat{\mathbf{r}}$$

where $\mathbf{M} = 5.972 \times 10^{24}$ kg is the mass of the earth and $\mathbf{G} = 6.674 \times 10^{-11}$ N·m²/kg² is the universal gravitational constant. Here $\hat{\mathbf{r}}$ is the position vector with its origin as earth's centre and its magnitude is $\mathbf{r}$ which is the sum of the mean radius of the earth (6378.14 km) and the orbital altitude or height of the satellite from the earth's surface.

Thus, we get the second-order vector ODE for $\mathbf{r}(t)$ as

$$d^2r/dt^2 = - (\mu/r^2) \, \hat{\mathbf{r}} \qquad (A.1)$$

μ is the product of G and M and it equals 3.986×10^{14} N · m²/kg. The –ve sign on the right-hand side indicates that the force is directed towards the centre of the earth.

The equation does not contain $\mathbf{m}$, the mass of the satellite and so the equation is the same whatever be the earth's satellite. The different solutions are entirely dependent on the initial conditions, namely at $\mathbf{t} = \mathbf{0}$, the position vector $\mathbf{r_0}$ and velocity vector $\mathbf{v_0}$

For a circular orbit, the magnitude of the acceleration is a constant and equals the centripetal acceleration of $\mathbf{v^2/r}$. Equating this to μ/r^2, we get

$$V_{circular} = \sqrt{(\mu/r)} \qquad (A.2)$$

The time period **T** of one orbit then works out to be

$$T = \sqrt{(4\pi^2 r^3/\mu)} \tag{A.3}$$

In this two-body motion problem, the conservation of the orbital angular momentum vector immediately leads to the motion being confined to a plane defined by **r x v** as the orbit's normal direction.

A satellite orbit can be described by following six Keplerian parameters:

- Semi-major axis - **a**
- Eccentricity - **e**
- Inclination – **i**
- Right ascension of ascending node - **Ω**
- True anomaly - **φ**
- Argument of perigee – **ω**

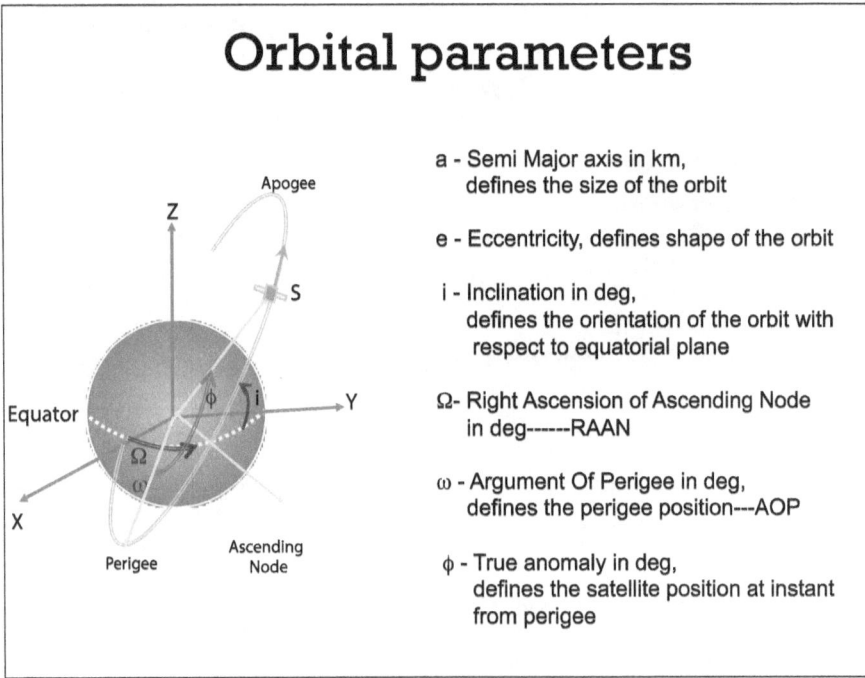

Orbital parameters

a - Semi Major axis in km, defines the size of the orbit

e - Eccentricity, defines shape of the orbit

i - Inclination in deg, defines the orientation of the orbit with respect to equatorial plane

Ω- Right Ascension of Ascending Node in deg------RAAN

ω - Argument Of Perigee in deg, defines the perigee position---AOP

φ - True anomaly in deg, defines the satellite position at instant from perigee

The semi-major axis gives the distance of the satellite's orbit from the centre of the earth and can be used for computing the orbital period.

The shape of the orbit is defined by eccentricity 'e' which can vary between 0 to 1. Most of the satellites (including Nanosatellites) placed in LEO and GEO orbits are circular where e = 0.

Inclination 'i' is defined as the included angle between the orbital plane and equatorial plane. The value of 'i' is in the range of 0-180 degrees. Generally, if the value is lesser than 90 degrees, the orbit is called prograde and if the value is greater than 90 degrees it is called retrograde.

The right ascension of the ascending node, RAAN (Ω) is defined as the angle between the vernal equinox and the point of ascending node on the equator.

True anomaly (φ) is defined as the angle subtended between the perigee and the satellite's location measured in the orbital plane.

The argument of perigee (ω) is the angle between the ascending node and the perigee.

These orbital parameters are pictorially shown in the figure below:

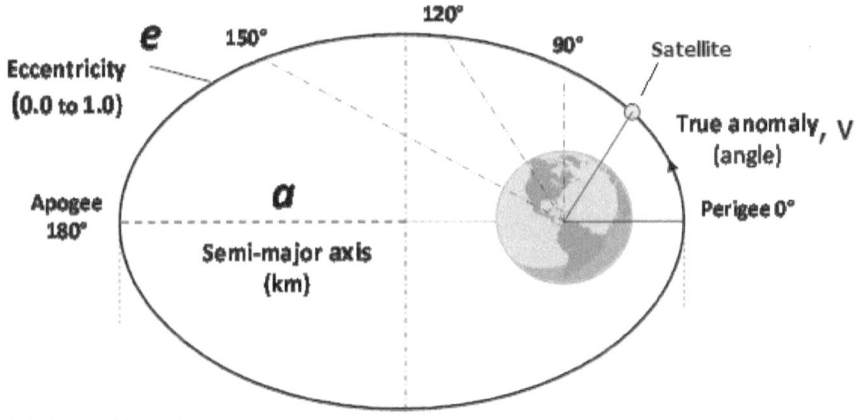

e defines ellipse shape
a defines ellipse size
V defines satellite angle from perigee

▲ Fig. A1.1 Orbit shape parameters (Courtesy: https://slideplayer.com/)

ANNEXURE - A2
PRINCIPLES OF REMOTE SENSING

Remote sensing (RS) is the technology of obtaining information from objects without being in physical contact with those objects. The human body has several remote sensing organs such as the eyes, ears, nose – but in relevance to earth observations, remote sensing may be defined as a collection of electromagnetic (EM) radiation reflected or emitted by various earth features such as soil, water, trees as well as buildings and by a set of payload instruments. The primary source of illumination for all the features on earth is the sun. The solar radiation passes through earth's atmosphere before reaching the surface thereby changing its intensity due to absorption, scattering and transmission by the constituents of the atmosphere (ozone, various gases, water vapour and particulate matter).

A typical intensity *vs.* wavelength curve of solar radiation is shown in **Fig. A2-1.** It can be seen that there are certain 'spectral windows' of wavelengths which are transmitted through the atmosphere and the remote sensing payloads are operated to collect radiation in these windows. While the radiation in the spectral range of 0.4 to 0.7 micrometres, known as visible light, is used for optical cameras, the range of wavelengths beyond 0.7 up to about 15 micrometres is used for infrared cameras. EM radiation up to about 3 micrometres is due to the reflection of solar radiation by the earth's surface while the radiation beyond 3 micrometres is primarily due to emission by the earth's surface because of its temperature (~300K).

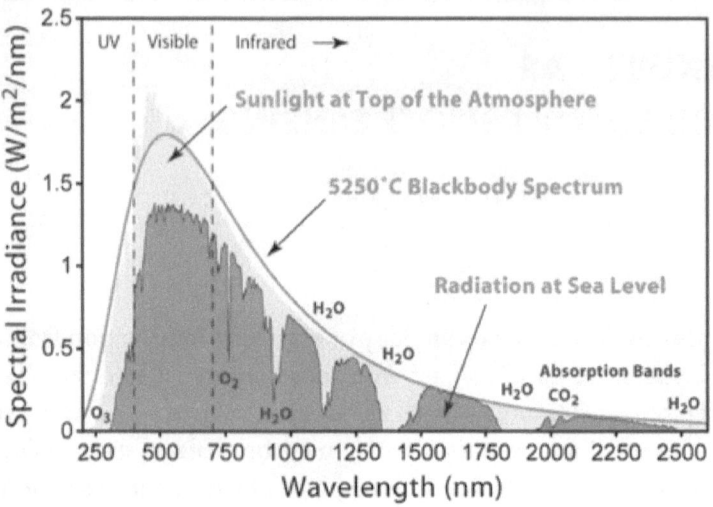

▲ **Fig. A2.1:** Solar radiation spectrum
(Courtesy: https://sunwindsolar.com/wp-content/uploads/2013/09/solar_spectrum.gif)

All the features on earth have their distinct characteristics in terms of reflection of solar radiation falling on them as shown in **Fig. A2-2.** From this figure, it can be seen:

- Water is a low reflecting feature since it absorbs most of the radiation falling on it – this is the reason why water bodies such as lakes, rivers and oceans look black in images.
- Vegetation absorbs blue and red light but reflects the green wavelength – the reason for the green appearance of all trees. But there is a sharp rise in reflection by leaves near the infrared region and this phenomenon is dominated by the moisture content on the leaves.
- Soil reflectance increases with wavelength but its overall reflectance reduces with increase in soil moisture

Both clouds and snow have a high reflectance in a visible range of wavelengths and appear 'white' in all images. However, in the short-wave infrared band (1.5 to 1.7 micrometres), the reflectance of snow is much lower than clouds. Due to the above mentioned characteristic signatures of various features, it is possible to identify/classify these earth features from remote sensing imagery. However, the accuracy with which these features can be identified/classified mainly depends on 'four resolutions' – spatial, spectral, radiometric and temporal resolution of the payload system.

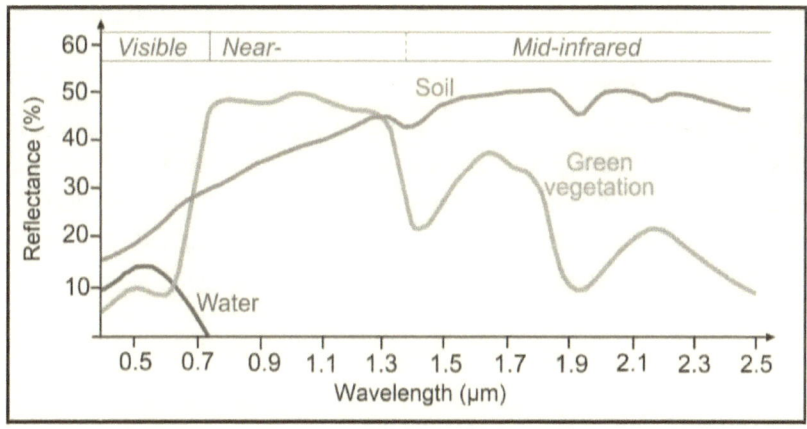

▲ **Fig. A2.2:** Spectral signatures of earth features (Courtesy: http://eumetrain.org/data/4/461/navmenu.php?tab=5&page=2.0.0)

- **Spatial resolution** is the minimum size of the object which can be distinguished in the image. For example, in an image with a resolution of one-meter even a motor car can be identified whereas, in a 30-meter resolution image, certain agricultural crops can only be identified. The spatial resolution of a camera can be derived from the simple equation **R = (H x d)/f** where R is the spatial resolution in metres, H is the orbit altitude in metres, d is the detector pixel dimension in metres and f is the focal length of the camera optics in metres. Thus, a camera with 100 mm focal length and five micrometres pixel size will produce an image of 30 metres resolution from an orbit of 600 kms altitude.
- **The spectral resolution** is the minimum bandwidth of the EM spectral range of wavelengths employed in the payload camera. For example, in a panchromatic camera, the spectral bandwidth will be about 450 nm (450 to 900 nm) while in a multispectral camera the bandwidth will be about 40 to 70 nm and in a spectrometer where there will be continuous spectrum (VIBGYOR), spectral bandwidth will be few nanometres.
- **Radiometric resolution** is the minimum detectable difference of the radiance values in the image. This is generally expressed in terms of the number of quantisation levels that each pixel intensity is converted into. In all the digital cameras, the analogue output of each pixel is converted into digital data (using A-to-D converter) and the number of bits of the ADC represents the radiometric resolution. Thus, a 10-bit data will have 1024 grey levels for each pixel while a seven-bit data will have only 128 grey

levels. Though it is preferable to select a higher number of quantisation bits, it should be kept in mind that the system noise should be less than the least significant bit (LSB).

- **Temporal resolution** is the minimum gap of time between two consecutive observations of a given area of interest. This depends on the demand of the application. For example, a disaster event such as a flood or an earthquake needs to be observed more frequently (one or two days) whereas an agricultural crop can be observed once in about 25 days.

ANNEXURE – A3
DIGITAL CODING TECHNIQUES

Aim of coding:
- To overcome errors in digital communication
- To improve communication reliability and performance
- Efficient utilisation of power and bandwidth

Criteria for the selection of coding techniques
- Type of communication channels and error
- Bit error rate (BER)
- Signal to noise ratio (SNR)
- Coding gain
- Code rate
- Bandwidth/data rate
- Data format
- Link margin
- Implementation design complexity of encoder and decoder

Types of error control coding are shown in **Fig. A3.1**
Codes are classified as block codes and tree codes. In block codes, information is divided into blocks and each block is coded independently. In tree codes, information is continuously coded like a tree. Block codes are the most suitable for the telecommand system and tree codes are used for the telemetry system.

The block diagram of a convolutional encoder is shown in **Fig.A3.2**

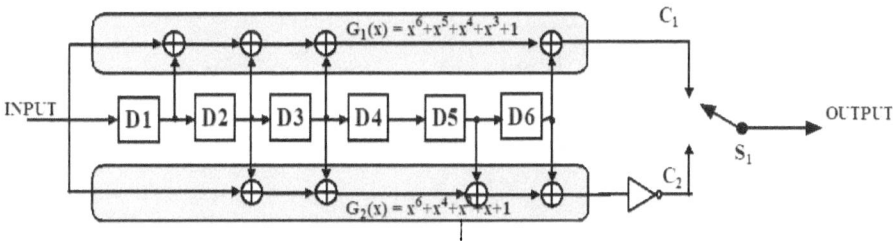

▲ **Fig. A3.2:** Convolutional encoder

A convolution encoder where for every input bit, two output bits are generated is called rate ½ encoder.

- $\boxed{D1}$ Represents a single bit delay

- $\bigoplus$ Represents Modulo-2 adder

- $\rightarrow\!\!\!\triangleright\!\!\!-$ Represents the Inverter

- For every input bit, two symbols are generated by the completion of a cycle for S1: position 1, position 2.
- S1 is in the position shown (1) for the first symbol associated with an incoming bit.
- The output symbol sequence is: C_1 (1), $C_2(1)$, $C_1(2)$, $C_2(2)$

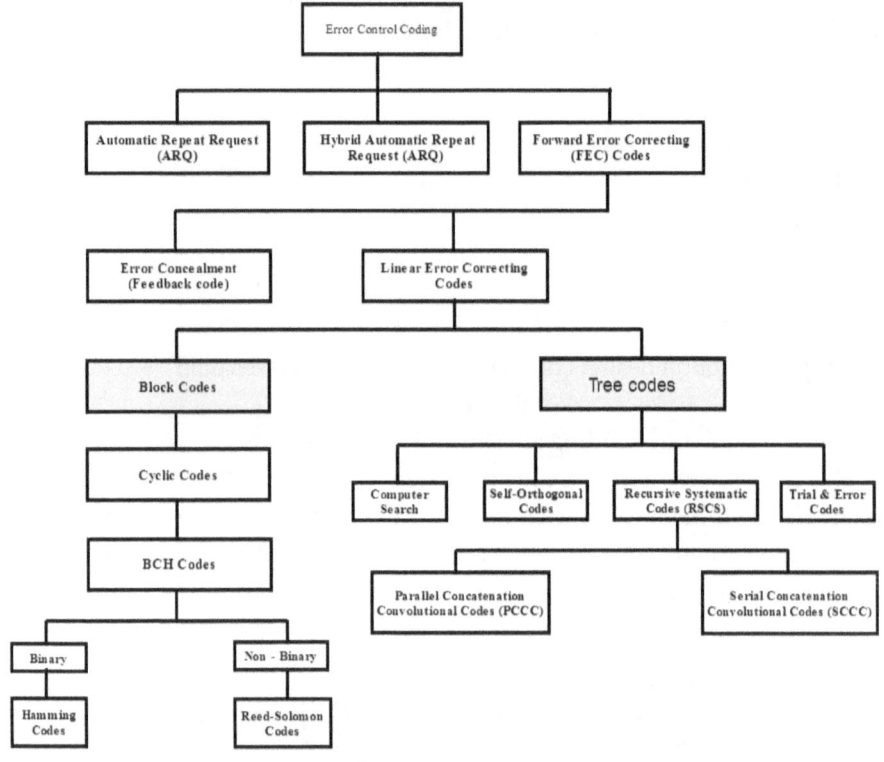

▲ **Fig. A3.1:** Types of error control coding

ANNEXURE - A4
FREQUENCY BANDS FOR NANOSATELLITES

Available amateur frequency bands for radio communication service category and the frequency coordination requirements for Nanosatellite missions.

▼ **Table. A4.1:** Amateur frequency bands and their applicability (Courtesy: ITU):

Frequencψ banδ (κHζ)	Metriχ reference	Frequencψ banδ (κHζ) (P = Regiov)	Applications
7,000-7,100	40 m	All regions	The HF bands are used only for limited amateur applications because of potential interference to and from terrestrial uses
14,000-14,250	20 m	All regions	
18,068-18,168	17 m	All regions	
21000-21450	15 m	All regions	
24,890-24,990	12 m	All regions	
Frequencψ banδ (MHζ)		**Frequencψ banδ (MHζ)**	
28-29.7	10 m	All regions	This band is used primarily in conjunction with an input or output in the 144 MHz band
144-146	2 m	All regions	These bands are in heavy use by numerous amateurs for inputs and outputs
435-438	70 cm	All regions, RR No. 5.282	
1260-1270	23 cm	All regions, RR No. 5.282 earth-to-space only	These bands are used as alternatives to the 144 MHz and 435 MHz bands because of congestion
2400-2450	13 cm	All regions, RR No. 5.282	
3400-3410	9 cm	Regions 2 and 3 only RR No. 5.282	
5650-5670	5 cm	All regions, RR No. 5.282 earth-to-space only	These bands are used by experimental amateurs
5830-5850		All regions secondary space-to-earth only	
Frequencψ banδ (ΓHζ)		**Frequencψ banδ (ΓHζ)**	
10.45-10.5	3 cm	All regions secondary	These bands are used for amateur communications
24-24.05	1.2 cm	All regions	

Frequencψ banδ (κHζ)	Metriχ reference	Frequencψ banδ (κHζ) (P = Regiov)	Applications
47-47.2	6 mm	All regions	These bands are used by experimental amateurs
76-77.5	4 mm	All regions secondary	
77.5-78		All regions primary	
78-81		All regions secondary	
134-136	2 mm	All regions primary	
136-141	2 mm	All regions secondary	
241-248	1 mm	All regions secondary	
248-250	1 mm	All regions primary	

▼ **Table. A4.2:** Radio communication services and their coordination requirements (Courtesy: Report ITU-R SA.2312-0 [09/2014])

Mission category	Typical radio communication service	Typical coordination requirement under Radio Regulations section II of Article 9* (for NGSO)
Educational and amateur radio	Amateur	Not subjected to coordination
Experimental, research	Space operation, space research, earth exploration	Not subjected to coordination
Commercial	Any radio communication service	May be subjected to coordination
*	All mission categories require advance publication and notification	

ANNEXURE – A5
RF LINK CALCULATIONS

In a satellite communication system, it is necessary to transmit/receive data within acceptable margins of loss of information, quantified in terms of bit error rate (BER). This depends on the quality of systems and the end-to-end link. The elements in a satellite ground station link are shown in **Fig. A5.1**. The link analysis/ budgeting is a way of quantifying the link performance. It consists of calculation and tabulation of the useful signal power and interfering noise power available with the receivers of both the satellite and the ground station. It is a balance sheet of gains and losses. The link analysis covers the entire communication path from the information source to the information sink. A measure of the satellite link performance is the ratio of carrier power to the noise power, denoted by C/N_0 at the receiver input. Link analysis is carried out both for uplink and downlink to estimate the received/available C/N_0 in analogue systems (or energy per bit to noise density $[E_b/N_0]$), which is the principal quality indicator in digital systems and comparing it with the minimum acceptable/ required C/N_0 (or E_b/N_0) for a given BER performance to establish the margin available after accounting for all the gains and losses in the link. The margin should be at least 6dB and3dB for a reliable TTC and payload link respectively. Link analysis/budgeting exercise also helps in visualising the overall system design and optimising the system configuration such as antenna size, power amplifier (SSPA) size, reduction of line losses and system noise temperature.

Data required for calculating the link margin for a Nanosatellite is detailed below:
1. Station longitude and latitude and height from mean the sea level
2. Maximum distance of satellite at an elevation angle of 5 degrees for path loss calculation
3. Uplink/ downlink frequency
4. Transmitter power - Pt
5. Gain of transmit antenna - Gt
6. Gain of receive antenna - Gr

7. Path loss - Ls
8. System noise temperature -T
9. Other losses - pointing, atmospheric, polarisation, Doppler shift in frequency- Lo
10. Onboard receive antenna gain
11. EIRP of an onboard antenna
12. Receiver antenna gain in the earth station
13. LNA noise temperature.
14. Data rate
15. Acceptable BER
16. C/No Uplink and C/No, Eb/No downlink

- Transmit power
- TX antenna gain
- Path losses
  - Free space
  - TX/RX antenna losses
  - Environmental losses
- RX antenna gain
- RX properties
  - Noise temperature
  - Sensitivity (S/N and ROC)
- Design margins required to guarantee certain reliability

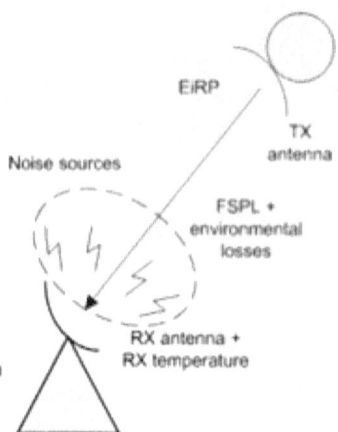

Note: satellite signals are usually very weak – requires careful link budget planning

▲ **Fig. A5.1:** Elements of the space communication link (Courtesy: internet)

The general equation for Link analysis is given below

$$(C/N_0) = P_t+G_t+G_r+L_s+L_{\theta t}+L_{\theta r}+L_a+L_l-10\log_{10}(k)-10\log_{10}(T_s) \text{ dB-Hz}$$

$$(E_b/N_0) = P_t+G_t+G_r+L_s+L_{\theta t}+L_{\theta r}+L_a+L_l-10\log_{10}(k)-10\log_{10}(T_s)-10\text{Log}_{10}(R) \text{ dB}$$

where C/N_0	:	Carrier power to noise density ratio
E_b/N_0	:	Energy per bit to noise density ratio
P_t	:	Transmitter power (dBw)
G_t, G_r	:	transmit and receive antenna gain (dBi)

L_s	:	Path loss or free space loss (dB)
$L_{\theta t}, L_{\theta r}$	:	Antenna pointing losses (dB)
L_a	:	Atmospheric absorption Loss (dB)
L_l	:	Line/ cable losses (dB)
T_s	:	System noise temperature (K)
k	:	Boltzmann constant (=1.38×10^{-23} J/K)
R	:	Data bit rate(bits/Hz)

These equations can be used without considering parameters not affecting the link (e.g., Pointing losses as beam widths are very wide and atmospheric losses)

Sometimes the receiver antenna gain and receive system noise temperatures are given as a combined parameter called G/T, the figure of merit of the receive system.

The final link margin for given BER performance is estimated as the difference between the available C/N_0 (or available E_b/N_0) and the required C/N_0 (or required E_b/N_0).

E.g.: Safety/ Link Margin: $(E_b/N_0)_{Available} - (E_b/N_0)_{Required}$

Additional equations useful for link calculation:
✓ EIRP = Transmit Rf power × Antenna transmit gain = $P_t \times G_t$ watts - is called equivalent isotropic radiated power
✓ Flux density = EIRP/$4\pi d^2$, where d = distance of the satellite from earth station in kilometres
✓ Antenna half-power beam width = $70\lambda/D$ in deg
✓ Antenna figure of merit = G/T in dB/ K where G is gain in dB and T is system noise temperature in Kelvin
✓ Flux density = F = $P_t/4\pi d^2$ where P_t = transmit power, d = distance of the receiver from transmit station
✓ Flux density for directional antenna = $P_t \times G_t/4\pi d^2$ = EIRP/$4\pi d^2$
✓ Let A_e, effective antenna aperture area = ηA_p, where A_p= physical aperture area, η = Aperture efficiency, G_r = gain of the receiving antenna = $4\pi A_e/\lambda^2$

Therefore Pr= $P_t \times G_t \times G_r [\lambda/4\pi d]^2$ = EIRP x G_r/ path loss
where path loss = $[4\pi d/\lambda]^2$
✓ Pr = EIRP + G_r - path loss in dBW
✓ Path loss can also be expressed by frequency

Path loss = 32.5+20 log d+20 log f

I.   Noise power= P_n=KTB, where K is Boltzmann constant= 1.38 X 10^{-23} J/K=-228.6 dBW/K/HZ, T= Noise temp, B= Bandwidth

II. Noise power spectral density= N_0= KT watts/HZ

III. P_r/N_0= C/N_0

IV. E_b= energy per bit= P_r / R_b R_b= bit rate.

V. E_b/N_0=C/N_0-R_b IN dB or C/N_0=E_b/N_0+R_b

ANNEXURE - A6
ANTENNA SYSTEM FUNDAMENTALS

An antenna system works on the principle of radio wave transmission. Radio waves are part of the electromagnetic spectrum starting from a frequency of 3 kHz to 300 GHZ. Space communications take place from the VHF frequency to the microwave frequency. Nanosatellites in particular use both VHF, UHF and microwave frequencies for communication.

A **transmitting antenna** converts electrical signals into electromagnetic waves and radiates them.

A **receiving antenna** converts electromagnetic waves from the received beam into electrical signals.

In two-way communication, a common antenna is generally used for both transmission and reception, unless the uplink and downlink frequencies are far apart, where two separate antennae are used.

Directivity of antenna

"The ratio of maximum radiation intensity of the subject antenna to the radiation intensity of an isotropic or reference antenna, radiating the same total power, is called the **directivity**." Its **radiation intensity** is focused in a particular direction while it is transmitting or receiving and hence the antenna is said to have its **directivity** in that particular direction.

An isotropic antenna is a theoretical antenna with a zero diameter radiating in free space in all directions with a spherical radiation pattern. All other antenna gains are referenced with respect to this antenna and are expressed as dBi.

Aperture efficiency of an antenna is the ratio of the effective radiating area (or effective area) to the physical area of the aperture. This radiation should be effective with minimum losses. The physical area (A_p) of the aperture is always higher than the effective aperture area (A_{eff}) and the ratio of effective aperture area and physical area defines the aperture efficiency (A_{eff}/A_p) - it is also called illumination efficiency.

Antenna efficiency is the ratio of the radiated power of the antenna to the input power accepted by the antenna. An Antenna is meant to radiate power given at its input with minimum losses. The efficiency of an antenna explains how much an antenna is able to deliver its output effectively with minimum losses in the transmission line. This is otherwise called as the **radiation efficiency factor** of the antenna.

Antenna gain, according to the standard definition, is the ratio of the radiation intensity in a given direction to the radiation intensity that would be obtained if the power accepted by the antenna were radiated isotropically.

The term **antenna gain** describes how much power is transmitted in the direction of peak radiation to that of an isotropic source. **The gain** is usually measured in **dB**. Unlike directivity, antenna gain takes the losses also into account and hence focuses on the efficiency.

VOLTAGE STANDING WAVE RATIO (VSWR) AND REFLECTED POWER

The standard definition of VSWR is "The ratio of the maximum voltage to the minimum voltage in a standing wave, in a transmission line."

Bandwidth is the band of frequencies between the higher and lower frequencies over which a signal is transmitted.

Feed system functions in two ways. It converts the radio frequency currents from the transmitter to radio waves as input to the antenna which in turn converts them into high energy beams and radiates. On the receiver path/direction, it collects the incoming low energy radio waves and converts them into electric currents and inputs them into the receiver.

The **Doppler effect** (or the **Doppler shift**) is the change in frequency or wavelength of a wave for an observer who is moving relative to the wave source. The received frequency, f_r is given by (assuming $v \ll c$),

Received frequency,

$$f_r = (1 + v/c)f_0$$

and the frequency difference due to Doppler shift,

$$\Delta f = (f_r - f_0) = (v/c)f_0$$

Where c= speed of light, v= relative velocity between observer & source (measured along the line of sight of v) - v is positive when the line of sight distance between the observer and the source is decreasing & negative when the line of sight distance is increasing, f_0= transmitted frequency

Yagi-Uda antenna: This is a very simple high gain antenna with high directivity and very easy to construct and test. The frequency range in which the Yagi-Uda antennae operate is **30 MHz to 3GHz.** It is mounted on a tall mast just above the equipment room. The antenna can be moved in both azimuth and elevation by the tracking system consisting of elevation and azimuth motors to exactly follow the path of a satellite.

The antenna can be constructed to match the polarisation of both the transmitter and receiver signals. The polarisation can be linear or circular and right or left circular (RHCP AND LHCP). The antenna has four important components.

Yagi-Uda Antennas

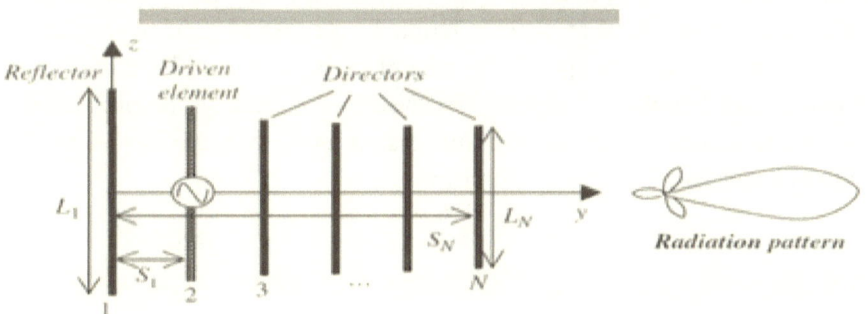

▲ **Fig. A6.1:** Yagi-Uda antenna (Courtesy: https://slideplayer.com/)

Driven element: The driven element in the Yagi antenna is the element to which power is applied. It is normally a half-wave dipole or often a folded dipole. The dipole has a torus type of radiation pattern.

Reflector: The reflector is an element suitably placed behind the driven element to make the antenna radiation directional in the direction of transmission. The Yagi antenna will generally have only one reflector. The length of the reflector is generally more than the driven element, which is a half-wave dipole. The reflector will add around 4 or 5 dB of gain in the forward direction.

Director: The director or directors are placed in front of the driven element, i.e. in the direction of maximum sensitivity. The spacing of directors depends on the wavelength and the number of directors on the mast is a trade-off between the additional gain and the increase in complexity. The maximum gain a single Yagi can attain is 18-20dBi.

The reflector is 5 per cent longer than the dipole - driven element and the main director is 5 per cent shorter than the dipole and located 0.25 λ in front of the dipole. The directors are made progressively shorter to make it taper in the direction of the transmitting antenna.

Cross Yagi antenna: The cross Yagi basically consists of two interleaved Yagi-Uda arrays with each of the arrays perpendicular to the other. Using this setup, while ensuring a proper phase-delay between the two arrays, a circular polarisation can be achieved.

Polarisation: The direction of the line traced by the tip of an electric vector is defined as the polarisation of the wave. Polarisation can be linear or circular. The transmission and reception frequency polarisation may be different. Normally, LEO ground antenna with circular polarisation is preferred as the onboard antenna will be an omni antenna. In the case of circular polarisation, the transmission and reception can take place both in the LHCP and RHCP. When a linearly polarised signal is received by a circularly polarised antenna then there will be a 3dB loss in transmission, even then this is preferred for LEO to avoid the signal from fading away.

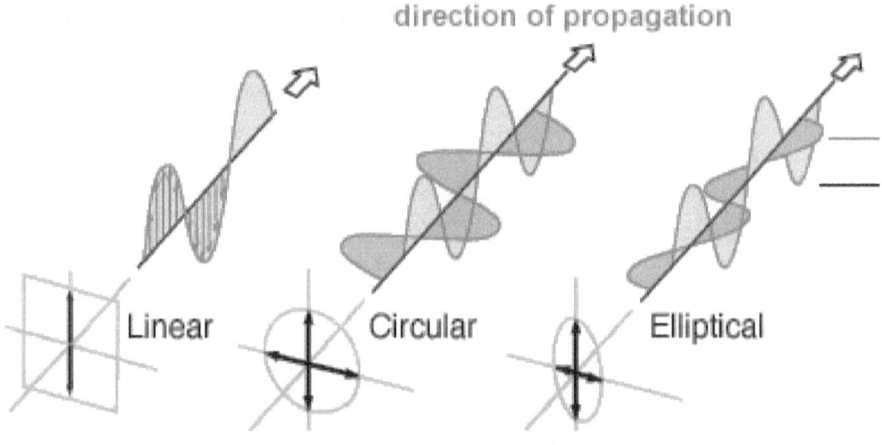

▲ **Fig. A6.2:** Polarisation definition

Antenna radiation pattern: A typical antenna radiation pattern is shown below. The main lobe and its beam width decide the gain of an antenna. A sharper and narrower beam indicates high directivity and gain. A good antenna will have low sidelobe and low back lobe.

Half power beam width is specified as the angle θ subtended by the -3dB lines on either side of the beam centre. The geometrical axis of the antenna and the RF beam axis theoretically should be the same for best pointing.

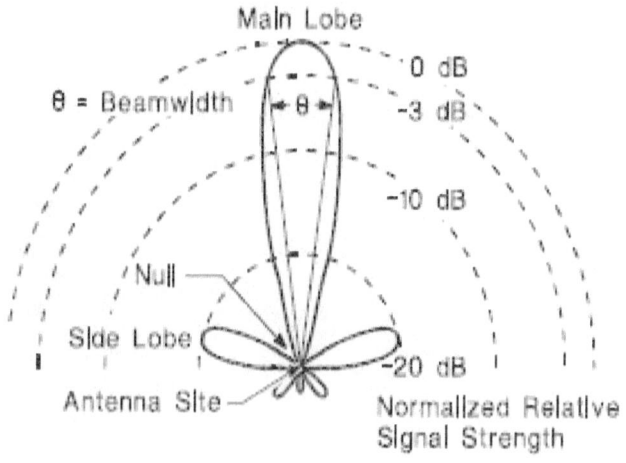

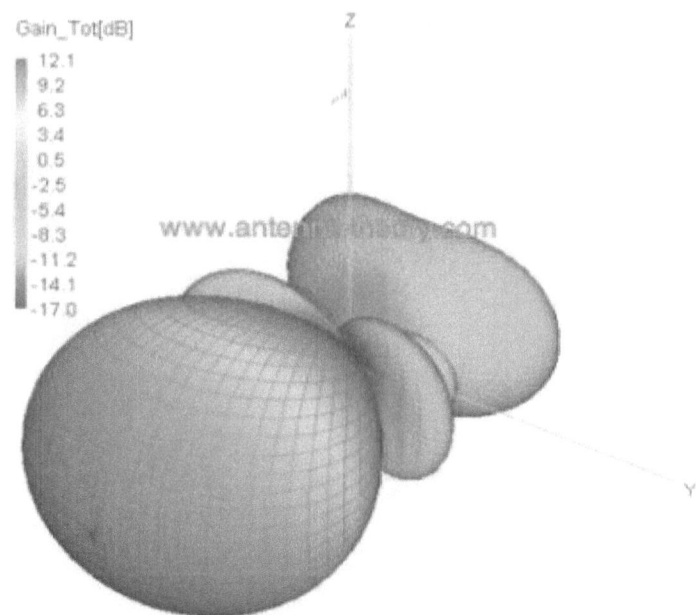

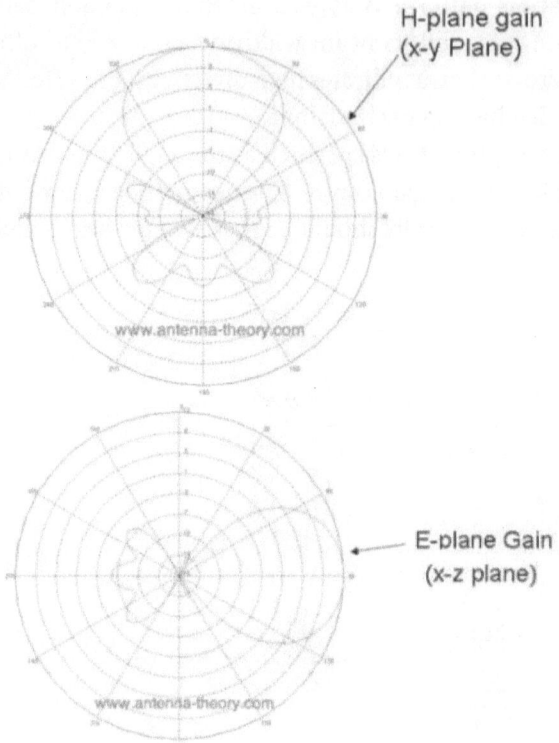

H-plane gain
(x-y Plane)

E-plane Gain
(x-z plane)

(Courtesy: www.antenna-theory.com)

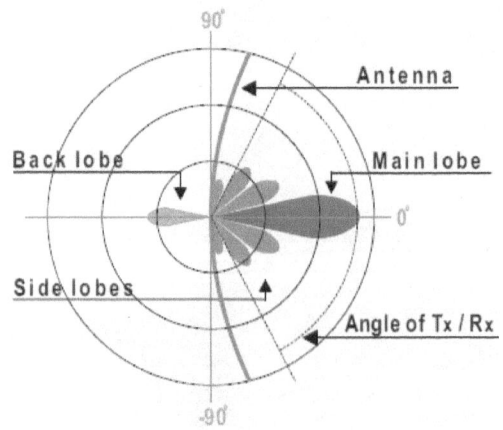

(Courtesy: www.racom.eu)

▲ **Fig. A6.3:** Antenna Pattern of Typical Yagi-Uda Antenna.

ANNEXURE - A7
EARTH STATION TESTING

Earth stations need to be tested/evaluated for several important/critical parameters before they are declared operational. These tests are generally categorized as i) RF tests; ii) Antenna tests and iii) Tracking/servo tests.

End-to-end satellite clearing tracking and data reception are the ultimate proof of performance test for an earth station.

RF Checks: G/T, system noise temperature, EIRP, harmonics/spurious, station cable losses, BER threshold, lock/track sensitivity, Doppler compensation and Doppler rate capability, blockage and noise survey, station software validation, station frequency and timing system checks as well as receive and transmit chain isolation.

Servo systems checks: Velocity, acceleration, pointing, angle measurements, and tracking in various modes/commanding/data reception.

Antenna safety checks: Hardware, software and mechanical limit checks, stow lock checks and emergency stop checks

ANTENNA TESTS

Radiation pattern (far-field antenna radiation pattern measurement): The radiation pattern is the representation of the radiation characteristics of the antenna as a function of the azimuth and elevation angles for any particular frequency. The three-dimensional pattern is decomposed into two orthogonal E and H field planes and Z-axis being the line joining the transmitter and receiver antenna and is perpendicular to radiation apertures. The test antenna is in the receiving mode and the standard antenna is fed by a stable source. The received signal is measured using a receiver. The output of the receiver is fed to Y-axis of XY recorder and angle information fed X-axis input of the XY recorder. This will give amplitude versus angle plot of the radiation pattern.

Antenna drive system and mount-angle calibration: The antenna's azimuth angle encoder at zero degree has to be aligned with respect to true north. There are many methods of finding the local true north. The antenna drive unit will be able to drive the antenna ± 270 degrees in the Azimuth and ± 90 degrees in the elevation. The antenna's dynamics will be (velocity and acceleration) designed to handle the expected rates.

Antenna blockage

An RF blockage survey is conducted before installing the antenna for 360-degree azimuth rotation. It is ensured that there is no blockage in the antenna tracking direction above two degrees in elevation over the entire range of the azimuth angle.

Phase centre measurement: Phase centre is a particular point in the antenna geometry which represents the radiation centre.

Beam width: The beam width is calculated from the angle subtended by 3dB and 10dB points on both sides of the antenna pattern.

Gain measurements: There are three methods to measure the gain of an antenna:
1. Standard antenna method
2. Two antennae method
3. Three antennae method

The first method is the simplest. This involves a test antenna of known gain Gr in the receiving mode and received power Pr is recorded. Then test antenna is replaced by the standard antenna of Gain Gs and received power Ps is recorded. Then,

$$Pr/Ps = Gr/Gs$$
$$Gr \text{ in dB} = Gs \text{ in dB} + 10\log(Pr/Ps).$$

Important factors considered for measuring the gain are impedance mismatch, polarisation mismatch and reflection from other sources, all at the minimum. The antenna is aligned with boresight, radiation face to face. A proper measuring set up with calibrated equipment is used for measuring the gain and other parameters.

Directivity measurements

Directivity is measured from the radiation pattern in two principal planes E and H planes. ΘE and ΘH, the half-power beam width, in these planes are determined from the radiation pattern.

Polarisation checks

Polarisation checks are the most important for the transmission and receiving signals. These are determined by using the antenna in the transmitting mode and probing the polarisation by using a dipole antenna in the plane that contains the direction of the electric field. The dipole is rotated in the plane of polarisation and the received voltage pattern is recorded.

- Linear polarisation: The output voltage pattern is a figure of eight
- Circular polarisation: The output voltage pattern is circular

Tracking checks

The most important function of an earth station is its ability to track a satellite and receive data from it without interruption until the end of visibility. This can happen if the antenna is able to faithfully follow the satellite's path as per prediction. It should be able to acquire the signal as soon as the satellite emerges from the horizon. Therefore rigorous tracking exercises are conducted before the launch of the satellite. For tracking a Nanosatellite, a programme track is highly recommended as the feed that is used on Yagi-Uda antenna is not a tracking feed. Accurate predictions of the satellite path can be achieved with a two-line element programme. The antenna is used for tracking the existing satellite in the orbit - the received signal strength can be measured and the proper margin established (many such satellites are orbiting in the LEO in the same frequency range). The antenna can also be checked by using a boresight, which is located at a distance not less than $2D^2/\lambda$.

ANNEXURE – A8
NORAD TWO-LINE ELEMENTS (TLE) SET FORMAT

Six parameters/numbers are required to uniquely locate an object/satellite in an orbit in space at any given time with the epoch forming the seventh parameter/number). These six parameters/numbers are called orbital elements or Keplerian elements. Corrections applied to Keplerian elements, due to earth oblateness, atmospheric drag and gravitational effects of other celestial bodies are known as perturbations. A satellite's orbital elements change over time, so the elements are accurate only at the instant of the epoch. The six orbital elements/parameters are:

a. **Mean motion, N/semi-major axis, a** (defines the size of the orbit)
b. **Eccentricity, e** (defines the shape of the orbit)
c. **Inclination, i** (defines the orientation of the orbit with respect to the earth's equator)
d. **Right ascension of ascending node (RAAN), Ω** (defines the location of the ascending orbit point with respect to the vernal equinox)
e. **Argument of perigee, ω** (defines the low point, perigee, of the orbit w.r.t. the earth's surface)
f. **Mean anomaly, M/ true anomaly, v** (defines the satellite position in the orbit w.r.t. the perigee at a given time, epoch)

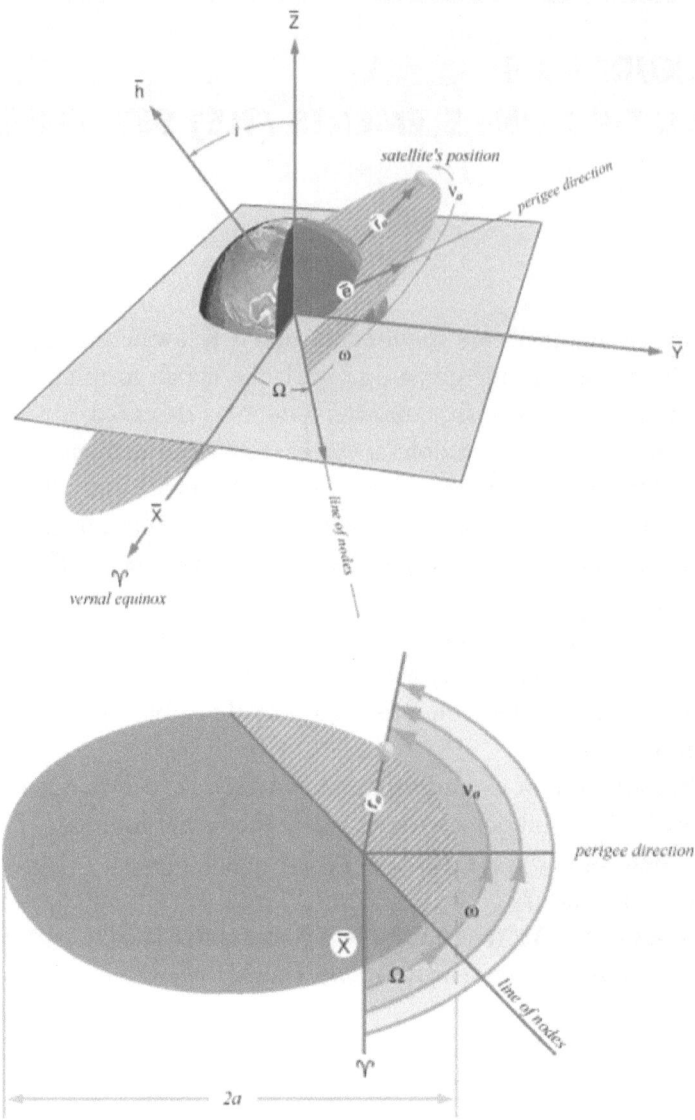

▲ **Fig. A8.1:** Orbital elements - visual description
(Courtesy: https://www.daviddarling.info/)

The orbital parameters of a satellite can be summarised into two lines of information with a maximum of 69 alphanumeric characters called "two-line element (TLE)" set. The TLE set adopts mean anomaly and mean motion instead of the true anomaly and semi-major axis respectively in its format. The accuracy is 1 km at the specified epoch time and then grows at 1km to 3 kms per day.

A TLE set is a data format encoding a list of orbital elements of an earth-orbiting object/satellite for a given point in time, the epoch.

The United States Air Force tracks all detectable objects/satellites in the earth's orbit, creating a corresponding TLE set for each object/satellite and makes available TLEs for non-classified objects on the internet at space-track. org or CelesTrak.

With suitable prediction formula/propagator, using TLE as data input, a satellite's state (position and velocity) at any point in the past or future can be estimated to some accuracy. The TLE data representation is specific to the simplified perturbations models (SGP, SGP4, SDP4, SGP8 and SDP8), so any algorithm using a TLE as a data source must implement one of the SGP propagator models to correctly compute the state at a time of interest. The TLE data set also includes the first and second time derivatives of mean motion and aerodynamic drag terms for use in simplified general perturbations (SGP) propagation model. TLEs can describe the trajectories only of earth-orbiting objects/satellites.

These two lines which are referred as "line 1" and "line 2" are preceded by a "line 0" also containing 69 characters but of which only 24 can serve to describe the space object/satellite name. The title (line 0) is not required as each data line includes a unique object/satellite identifier code. TLE sets are provided in an ASCII formatted text file

For a better understanding of the NORAD TLE set, TLE of International Space Station (ISS) and description of each line and column are given below:

```
ISS (ZARYA)
1 25544U 98067A 04236.56031392.00020137 00000-0 16538-3 0 9993
2 25544 51.6335 344.7760 0007976 126.2523 325.9359 15.70406856 328903
1234567890123456789012345678901234567890123456789012345678901234567890
ref. Line number
1 2 3 4 5 6 7
```

▲ **Fig. A8.2:** International Space Station TLE

▼ **Table. A8.1:** International Space Station TLE Description

Columns	Example	Description
Line 0		
1 - 24	ISS (ZARYA)	Object common name based on information from the satellite catalogue

Columns	Example	Description
Line 1		
1	1	Line number
3-7	25544	Satellite catalogue number
8	U	Elset classification
10-17	98067A	International designator
19-32	4236.560314	Element set epoch (UTC) * Note: Spaces are acceptable in columns 21 and 22
34-43	0.00020137	1st Derivative of the Mean Motion with respect to Time
45-52	00000-0	2nd Derivative of the mean motion with respect to Time (decimal point assumed)
54-61	16538-3	B* drag term
63	0	Element set type
65-68	999	Element number
69	3	Checksum
Line 2		
1	2	Line number
3-7	25544	Satellite catalogue number
9-16	51.6335	Orbit inclination (degrees)
18-25	344.776	Right ascension of ascending node (degrees)
27-33	7976	Eccentricity (decimal point assumed)
35-42	126.2523	Argument of perigee (degrees)
44-51	325.9359	Mean anomaly (degrees)
53-63	15.70406856	Mean motion (revolutions/day)
64-68	32890	Revolution number at epoch
69	3	Checksum

Several commercial and open source/free orbit propagator software are available for different operating systems (Windows or Linux) that use TLE sets as input data for satellite tracking/antenna pointing.

REFERENCES

1. 'Satellite Tracking using NORAD TLE Set Format' by Emilian Ionut Croitoru, Gheorghe Oancea. Transilvania University of Brasov, Romania
2. www.celestrak.com/software/
3. www.daviddarling.info/encyclopedia/o/orbital_elements.html

ANNEXURE – A9
ADCS RELATED DEFINITIONS

ATTITUDE ACQUISITION

Attitude acquisition is the function of bringing the satellite from an unknown orientation to a known and desired orientation. The desired orientation can be either sun pointing or earth pointing depending on the mission's requirement. The last stage of the rocket injects the satellite into the orbit with some angular velocities. Essentially the satellite will be in tumbling mode, spinning about all the three axes with a low angular velocity of the order of few degrees per second. The ADCS first reduces these angular velocities using the magnetic torquer coils/propulsion thrusters. After the reduction of rates, the satellite will be in some random unknown orientation. The ADCS determines the satellite's current attitude with the help of sensors such as sun sensor, star sensor or earth sensor and generates control signals to the actuators such as MTC or thrusters. The actuators generate the required torque to orient the satellite in the desired direction.

ATTITUDE STABILISATION

After completing the attitude acquisition, the attitude of the satellite has to be maintained within tolerable limits for maximum power generation and also to keep the temperature of all the sub-systems within specified limits. The major environmental factors disturbing a satellite's attitude are solar radiation, aerodynamics and magnetic torques. Attitude stabilisation is an operation to maintain the desired orientation of the satellite in the face of these disturbances. The ADCS performs this function in a closed-loop manner controlling the attitude and the angular rates using the feedback from attitude sensors. There are two types of attitude stabilisation techniques: Passive and active. The passive technique does not draw power from the satellite and functions using the naturally available forces such as the gravity gradient and the magnetic field of the earth. The active technique uses the momentum/reaction wheels and/or propulsion thrusters and magnetic torquers to generate the required torques.

ATTITUDE DETERMINATION

Attitude determination is the process of combining available sensor inputs with knowledge of the spacecraft dynamics to provide an accurate and unique solution for the attitude state as a function of time, either onboard for immediate use or post facto after processing on the ground. Attitude determination exercise requires finding three independent quantities to represent the satellite's orientation. The output of this exercise provides the attitude estimate or solution, which is obtained by using sensors to relate information about external references, such as the stars, the sun, the earth or other celestial bodies with respect to the orientation of the spacecraft. Generally, any single sensor has a noise level or other drawbacks that prevent it from providing a fully satisfactory attitude solution at all times. Therefore, more than one sensor is often used to meet the mission requirements.

It is useful to divide attitude determination approaches into two categories. The first category comprises static determination approaches that depend on measurements taken at the same time or close enough in time so that the satellite's motion between the measurements can be ignored or easily compensated for. The second category comprises filtering approaches that make explicit use of knowledge of the motion of the satellite to accumulate a "memory" of past measurements. Static approaches require enough observations each time to fully compute the attitude but they typically don't require *a priori* attitude estimate. A purely deterministic approach incorporates just enough observation information to uniquely determine the attitude. A star tracker onboard a satellite observes line-of-sight vectors to several stars, which are compared with known inertial line-of-sight vectors from a star catalogue to determine the attitude of the satellite. In many satellite attitude determination methods, the attitude observations are naturally represented as unit vectors. Typical examples are the unit vectors giving the direction to the sun or a star and the unit vector in the direction of the earth's magnetic field.

ATTITUDE CONTROL

Attitude control is the combination of the prediction of and reaction to a satellite's rotational dynamics. Because the satellite exists in an environment of small and often highly predictable disturbances, they may in certain cases be passively controlled. A satellite may be designed in such a way that the environmental disturbances cause the satellite attitude to stabilise in the orientation needed to meet mission goals. Alternately, a satellite may include actuators that can be used to actively control its orientation. These two general types of attitude control are not mutually exclusive. A satellite may mostly be controlled passively and yet include actuators to adjust the attitude in small ways or to make attitude manoeuvres (i.e. slews) to meet other objectives, such as targets of opportunity or communication needs.

AGILITY

Remote sensing satellites require rapid reorientation capability. An agile satellite is much more efficient by way of increased image throughput. Agility is defined as how fast the satellite's orientation can be shifted. Higher agility can be achieved by configuring the satellite to be small and using high torque capacity actuators.

Pointing Control Definitions (see **Fig. A9-1**)

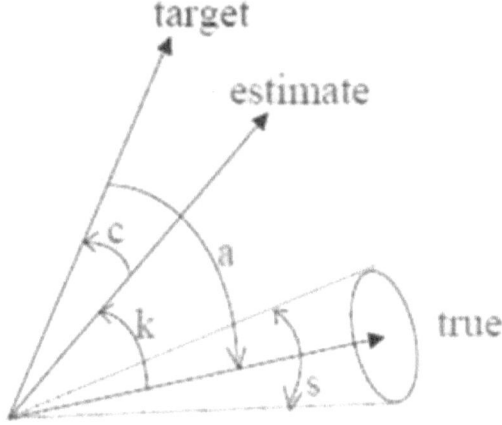

▲ **Fig. A9.1:** Attitude pointing definition

Target: Desired pointing direction, s: Stability (peak to peak motion); k: Error in attitude knowledge
True: Actual pointing direction, c: Control error
Estimate: Estimate of true (instantaneous), a: pointing accuracy

ACTUATOR SIZING

Momentum and torque requirements of reaction wheels
The main contributors to the system momentum and torque requirements are
1. Environmental disturbances
2. Attitude slew manoeuvres
3. Internal disturbances from the rotation of solar array, scanning mechanisms and antenna

To determine the necessary momentum capacity, one must distinguish between cyclic and secular disturbances in the satellite's environment. The momentum build-up due to cyclic torques averages out every half orbit or full orbit depending on whether the disturbance is periodic at an orbital frequency or twice the orbital frequency. The reaction wheels should be able to store the full cyclic component of momentum without the need for frequent momentum dumping. Therefore, the average disturbance torque for 1/4 or 1/2 an orbit determines the minimum capacity of the wheels. The secular component of the momentum will also need to be retained for the duration of the time the satellite must be operational without a momentum dump being performed. This time may be determined by requirements on payload observation continuity or it may be the amount of time the satellite must survive without ground intervention. In addition, any attitude manoeuvre for sun pointing, re-orientation for image capturing or recovery from safe mode needs additional storage momentum required for steering.

The maximum torque of the reaction wheel does not arise from the feeble environmental disturbances. Rather it depends on steering requirements for payload pointing or recovery from safe mode. The required slew time depends on the wheel torque capacity, current momentum bias and any attitude or rate limits that may be imposed. The reaction torques due to antenna and solar array tracking must also be considered while selecting and sizing the reaction wheels.

SIZING OF THE MAGNETIC TORQUERS

Magnetic torquers are used for the following functions:
1. Momentum de-saturation of reaction wheels/momentum wheels
2. Three axes attitude control
3. De-tumbling of the injection rates after the satellite's separation from the launch vehicle

The torque developed by magnetic torquers depends on the earth's magnetic field, which is a function of the orbital height and the dipole moment of the torquer. The torque developed must overcome all the peak disturbances as well as reduce the de-tumbling operation time. Nanosatellites generally do not include thrusters in the configuration and therefore the de-tumbling needs to be carried out using magnetic torquers. Generally, the torquer capacity is decided based on the de-tumbling requirements since during de-tumbling, power generation is very less and the entire satellite's power requirement is supported by the battery. De-tumbling should be completed before the battery goes below the safe level and hence would require a high capacity torquer. The other requirements of three axes control and momentum de-saturation will automatically be met.

SELECTION OF SENSORS

Sensor selection is directly influenced by the required orientation of the satellite (e.g. earth, sun or inertial-pointing) and its accuracy. Other influences include redundancy, noise characteristics, fault tolerance and field of view requirements. All the available candidate sensor suites are identified and a trade-off study conducted to determine the most cost-effective approach to meet the needs of the mission. The existence of off-the-shelf components and software can strongly affect the outcome.

An inertial reference unit (IRU) is normally used for attitude propagation. The IRU requires updates of the absolute attitude at frequent intervals in order to make use of the short-term stability of the IRU and the long-term advantages of star trackers and horizon sensors.

For an earth-pointing satellite, horizon sensors provide a direct measurement of pitch and roll axes. Depending on the accuracy required, we can use sun sensors, magnetometers or momentum-bias control with its roll-yaw coupling for the third degree of freedom. For inertial pointing spacecraft, star and sun sensors provide the most direct measurements and IRUs are ideally suited. Using the orbit parameters, it is possible to convert measurements in the reference frame to any other required reference frame. Either the orbit parameters are uplinked to the satellite from the ground tracking station and propagated by the onboard processing system or they are obtained from onboard GPS antennae. Attitude determination using magnetometer also requires orbit information to generate the reference earth's magnetic field.

ANNEXURE – A10
PROPULSION SYSTEMS FOR SMALL SATELLITES

Propulsion means to push forward or drive an object forward. A propulsion system is a machine that produces thrust to push or move an object forward. This chapter provides an introduction to various types of propulsion systems that can be deployed in satellites. Space propulsion, which so far has been exclusive to large and expensive satellite missions, is now becoming a realistic possibility for many small satellites.

The propulsion system in a satellite has two main functions: (i) Attitude control and (ii) Orbit acquisition and maintenance. A satellite's three axes attitude is mainly controlled with reaction wheels and whenever the wheel speeds reach a set limit due to transfer of angular momentum, the propulsion thrusters are fired to dump the momentum thereby reducing the wheel speeds. To achieve the intended orbit after injection from the launch vehicle, the satellite propulsion system is used. Also, once placed in their orbit, satellites are subjected to drag due to factors such as solar radiation pressure and earth's gravity, which reduce the orbital speed over time. So, to stay in the correct orbit for a longer period of time, some form of propulsion is occasionally necessary to make corrections for the drag induced orbital errors. A satellite's useful life usually ends once it has exhausted its ability to adjust its orbit.

Therefore, whatever may be the size and orbit of a satellite, some form of propulsion system will be required to control its orbit and attitude throughout its operational life. This requires the generation of thrust (F) to produce either translation or torque to achieve movement of the satellite with respect to its centre of mass (CM) as depicted in **Fig. A10**-1 **(a) and (b)** respectively. Thrust for such movements is generated using small rockets called thrusters.

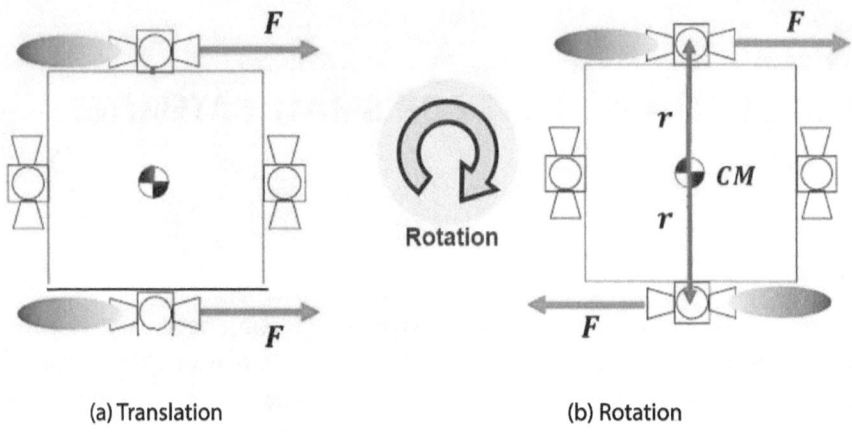

(a) Translation   (b) Rotation

▲ **Fig. A10.1:** Types of thrust required for a satellite

For near-earth satellites this is accomplished by one or more of the following techniques:

(**a**) Inert gas propulsion
(**b**) Monopropellant propulsion
(**c**) Bi-propellant propulsion
(**d**) Hybrid propulsion
(**e**) Electric propulsion

Among these (a), (b), (c) and (d) are classified as "chemical propulsion" because they use some kind of chemical agent to generate thrust. However, the basic principle of generating thrust using any of the above techniques remains the same.

THE BASIC ROCKET EQUATION

A basic rocket is schematically depicted in **Fig. A10**-2. It consists of a combustion or reaction chamber and a convergent-divergent nozzle.

All rocket engines produce thrust by accelerating a working fluid. Chemical rocket engines use the combustion of propellants to produce exhaust gases as the working fluid. The high pressures and temperatures of combustion are used to accelerate the exhaust gases through a convergent-divergent nozzle to produce thrust.

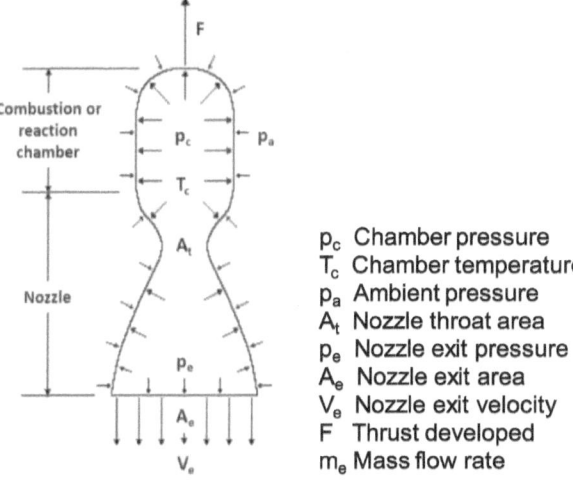

Combustion or
reaction
chamber

p_c

p_a

T_c

Nozzle

A_t

p_e

A_e

V_e

p_c Chamber pressure
T_c Chamber temperature
p_a Ambient pressure
A_t Nozzle throat area
p_e Nozzle exit pressure
A_e Nozzle exit area
V_e Nozzle exit velocity
F Thrust developed
m_e Mass flow rate

▲ **Fig. A10.2:** Basic rocket configuration

The thrust developed by a rocket engine is given by,

$$F = mV_e + \left(p_e - p_a \right) A_e \quad \dots\dots\dots\dots (1)$$

Where,
F is the thrust (N)
$\dot{m}$ is the mass flow rate of exhaust gases (kg/s)
V_a is the exhaust velocity at nozzle exit (m/s)
p_e is the pressure at nozzle exit (Pa)
p_a is the ambient pressure (Pa)
A_a is the nozzle exit area (m²)

The first term in equation (1) is called momentum thrust and the second term as pressure thrust. As can be readily seen, the pressure thrust part contains the ambient pressure, p_a. Since p_a decreases with altitude within the earth's atmosphere, the pressure thrust is applicable only for rocket engines of launch vehicles during their ascent phase and it becomes insignificant once the vehicle is outside the atmosphere. For the same reason, rocket thrusters used on satellites will have only momentum thrust and negligible pressure thrust. Therefore, for satellite thrusters, equation (1) reduces to

$$F = \dot{m}V_e \quad \dots\dots\dots\dots\dots (2)$$

An examination of equation (2) shows that a given thrust, F, can be achieved either with high mass flow rate and low exhaust velocity or low mass flow rate accompanied by high exhaust velocity. Here, the choice of propellant determines the best combination of mass flow rate and exhaust velocity.

DIFFERENCE BETWEEN THE LAUNCH VEHICLE AND SATELLITE PROPULSION

The purpose of a launch vehicle is to place a satellite precisely in the desired orbit. This requires work to be done against the gravitational pull of the earth and flight through a dense atmosphere before achieving the goal. A typical launch sequence takes about 15 to 40 minutes. Therefore, a launcher requires rocket engines to develop large thrust that can operate for several minutes. As an example, **Fig. A10**-3 **(a)** shows the Vulcain 2 engine used in the Ariane 5 launch vehicle. It produces a thrust of 1,359 kilo-Newtons (kN), operates for 605 seconds, measures 2.09 metres in diameter, is 3.44 metres long and weighs 1,800 Kgs.

A satellite, on the other hand, operates for several years. Since the space environment does not offer any significant resistance to movement, rocket engines of small thrust levels are adequate for operations in space. However, they are required to operate for several years reliably. As an example, **Fig. A10**-3 **(b)** shows a typical attitude control thruster used in satellites. It produces a thrust of 1N, operates in short pulses over several years, measures 50 mm in diameter, is 175 mm long and weighs 290 grams.

(a) Vulcain-2 Engine (Courtesy: Snecma)   (b) 1N Thruster (Courtesy: Arianespace)

▲ **Fig. A10.3:** Thrusters for (a) Launch vehicle (b) satellite

An efficient launch vehicle system requires a heavy satellite to be sent into orbit using as light a vehicle as possible and at a minimum cost. Similarly, an efficient satellite system requires a heavy a payload to be accommodated on as light a platform as possible at minimum cost.

Typical sources and categories of satellite mass are indicated in **Fig. A10-4**

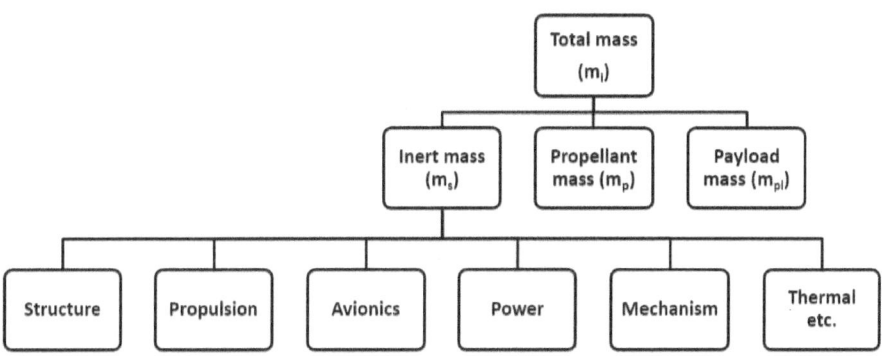

▲ **Fig. A10.4:** Sources and categories of space vehicle mass

Obviously for a given total mass, the inert mass and propellant mass should be low in order to maximise the payload mass. The influence of these different masses on a space mission can be assessed using the performance parameters indicated below:

(a) Mass ratio: Mass ratio (M_R) is defined as the ratio of final mass of a space system to its initial mass. It is generally considered as an efficiency indicator.

$$M_R = \frac{m_f}{m_i} \quad \dots\dots\dots\dots\dots\dots\dots \text{ (3)}$$

Where m_f is the final mass, which can be the mass of an empty vehicle plus the satellite mass for a launch vehicle or the useful payload mass plus the empty satellite mass for a satellite, and m_i is the initial mass, which can be the lift-off mass for a launch vehicle or the initial mass for a satellite.

In practice, however, essential peripherals such as structure, fuel, power and electronics are needed to accomplish a mission. Accounting for these, equation (3) can be written as,

$$M_R = \frac{m_f}{m_i} = \frac{m_{pl} + m_s}{m_f + m_p} = \frac{m_{pl} + m_s}{\left(m_{pl} + m_s\right) + m_p} \quad \dots\dots\dots\dots\dots \text{ (4)}$$

Where, m_{pl} is the useful payload mass, which can be the satellite mass for a launch vehicle or the payload mass for a satellite. m_s is the inert mass required to accomplish the mission, such as structure, power, propulsion hardware, avionics, power, mechanisms and thermal support. m_p is the propellant mass.

(b) Payload fraction: Payload fraction (λ) is defined as the ratio of useful payload mass to all the other mass required to support the payload. It is a common term used to characterise the efficiency of a particular design.

$$\lambda = \frac{m_{pl}}{m_i - m_{pl}} = \frac{m_{pl} + m_s}{\left[\left(m_{pl} + m_s\right) + m_p\right] - m_{pl}} = \frac{m_{pl}}{m_s + m_p} \quad \text{.................. (5)}$$

(c) Inert mass fraction: The inert mass fraction (δ) is the ratio of inert mass to the initial mass of a space system.

$$\delta = \frac{m_s}{m_i} = \frac{m_s}{m_s + m_p + m_{pl}} \quad \text{...................... (6)}$$

From the examination of equations (3), (5) and (6), it is apparent that for an efficient space system, the mass ratio (M_R) and payload fraction (λ) should be high and the inert mass fraction (δ) must be low. Further, from equation (4) it can be observed that the propellant mass (m_p) must be low for a given mass ratio which emphasises the use of high energy propellants.

PERFORMANCE FACTORS FOR A PROPULSION SYSTEM:

A propulsion system designer should consider the following four important factors while configuring a propulsion system for a satellite mission:

(a) Thrust, F, given by equation (1)

(b) Specific Impulse, I_{sp}, which indicates the efficiency of a propellant and given by

$$I_{sp} = \frac{F}{\dot{m}g_0} \quad \text{......................... (7)}$$

Where, g_0 is the acceleration due to gravity at sea level.

(c) Effective exhaust velocity, v_e, given by,

$$v_e = I_{sp}g_0 \quad \text{................................ (8)}$$

(d) Velocity increment, Δv, given by,

$$\ddot{A}v = v_e ln\left(\frac{m_i}{m_f}\right) = I_{sp}g_0\, ln\left(\frac{m_i}{m_f}\right) \ldots\ldots\ldots (9)$$

Equation (9) is known as the Tsiolkovsky rocket equation and it is used to estimate the achievable velocity increment for a given mass of propellant (which is $m_i - m_f$) or the propellant required for a given velocity increment.

PROPULSION SYSTEMS FOR SMALL SATELLITES

Although today's small satellites are physically small, they are nevertheless complex and exhibit virtually all the characteristics of a large satellite. Therefore, as an essential sub-system of a satellite, the propulsion system is also undergoing miniaturisation to accomplish its functions. In addition to the basic requirements of attitude and orbit control, propulsion systems are expected to deliver high performance and remain reliable throughout their mission. This should be easy to integrate into a satellite, service and maintain. These are the challenging demands of a small satellite propulsion system.

Since the majority of upcoming missions imply deployment of small satellites into LEO below 800 kms, their propulsion system must provide thrust sufficient for atmospheric drag compensation at low power consumption. Micro propulsion units based on the following technologies have either been used or demonstrated as suitable for small satellite applications:

- Cold gas propulsion (CGP) systems
- Liquid propulsion (LP) systems
- Solid rocket propulsion (SRP) systems
- Resistojets
- Radio frequency ion thrusters
- Hall effect thrusters
- Electro-spray thrusters
- Pulsed plasma thrusters (PPT)
- Solar sails

Following sections explain the principle of the operation of systems using the above technologies as well as description of a few applications.

COLD GAS PROPULSION (CGP) SYSTEM:

In this type of propulsion, liquid or gaseous propellants are ejected through a nozzle to generate thrust. The main components include propellant storage, flow control valve and a nozzle as shown **in Fig. A10-5**

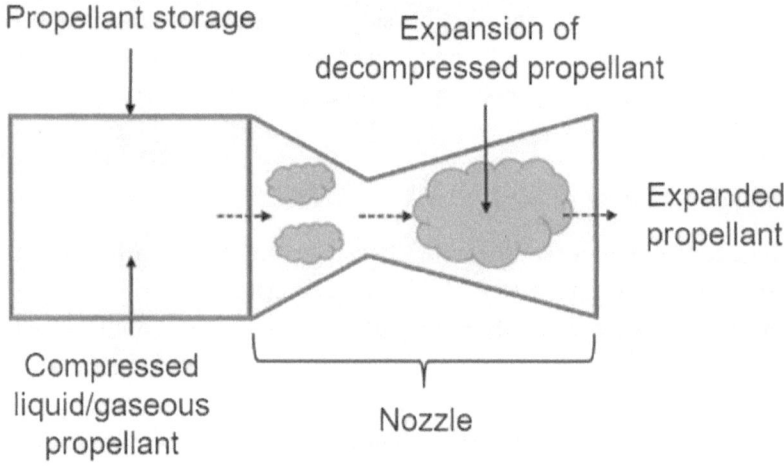

▲ **Fig. A10.5:** Cold gas propulsion system

The thrust produced is directly proportional to the pressure of the propellant stored and over the course of the mission, this pressure decreases (due to propellant usage) resulting in a decrease of the maximum thrust that is generated by the system. In a modified CGP system called solar thermal propulsion, concentrated solar energy is used to directly heat the propellant. As the propellant enters the nozzle at an elevated temperature, a significantly increased thrust and an increased specific impulse relative to a regular cold gas flow can be obtained. Cold gas propulsion systems have been used on CubeSat missions as can be observed from **Table. A10-1.**

▼ **Table. A10.1:** Examples of small satellites using cold gas propulsion

Company	Engine	Thrust (mN)	I_{sp} (Ns/kg)	Propellant	Heritage
SSTL, UK	SNAP 1	50	422	Liq. Butane	Giove A (600 kg) (2005-08)
UTIS-SFL, Canada	CNAPS	10-40	< 343	SF_6	CanX-4 & 5 (7 kg each) (PSLV C23 – 2014)
Microspace, Singapore	H1P1	1	422	Argon	POPSAT-H1P1 (3U/3.3 kg; 2014)
GOMSpace, Denmark	MEMS Cold gas	1	490-735	Methane	TW-1 (One 3U & two 2U) (Long March 11-2015)
VACCO Industries, USA	CPOD	25	392	R134a, R236fa	CPOD (3U) (Sch 2020) MarCO-A & B (6U/13.5 kg each) (2018)

As of now, cold gas propulsion has been successfully used for attitude and orbit control of CubeSat class of spacecraft including the interplanetary MarCO-A & B which were used for real-time data relay from Mars during the landing of InSight spacecraft on March 26, 2018.

Liquid propulsion (LP) systems
In this system, liquid propellants (either a single component or two components) are injected into a combustion chamber. The resulting products of combustion are ejected through a nozzle to generate thrust. A schematic diagram of a liquid propulsion system is shown in **Fig. A10-6.**

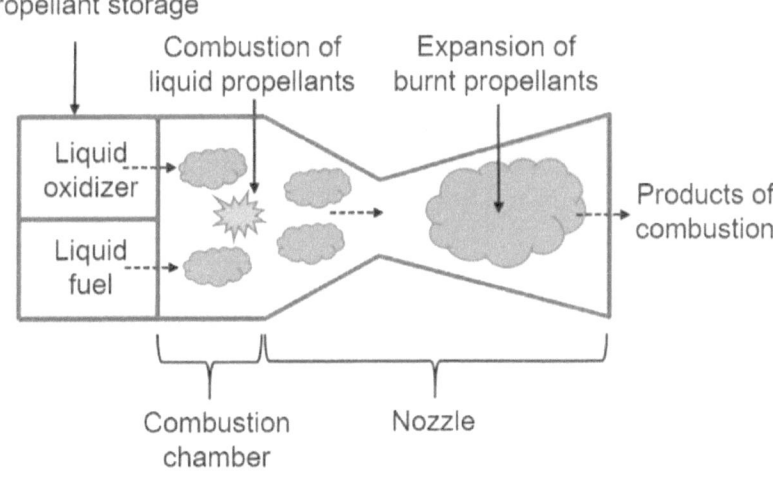

▲ **Fig. A10.6:** Liquid Propulsion System

The single-component propellant will use a catalyst to decompose while the two-component system may be self-igniting or use a separate igniter to initiate combustion. These propellants are generally toxic and difficult to handle. The performance depends on the propellant supply pressure. Of late, hydroxyl-ammonium nitrate (HAN) and ammonium dinitramide (ADN) based "green propellants" of low toxicity are gaining popularity to replace toxic propellants.

Liquid propulsion systems have been widely used on larger satellites. They have not been used on microsatellites so far due to difficulty in producing small thrusters. Technology to develop milli-Newton level liquid propellant thrusters is progressing at several places as shown in **Table. A10-2**.

▼ **Table. A10-2:** Examples of liquid propulsion systems

Company	Engine	Thrust (mN)	I_{sp} (Ns/kg)	Propellant	Heritage
Aerojet, USA	GPIM	400-1100	2305	HAN based	-
Aerojet, USA	MPS-120	260	2109	Hydrazine	-
Aerojet, USA	MPS-130	1.5	2354	HAN based	-
ECAPS, Sweden	HPGP	1000	2275	ADN based	PRISMA (180 kg)
Busek, USA	BGT-X1	100	2099	HAN based	-
Busek, USA	BGT-X5	500	> 2158	HAN based	-
Tethers Unlimited, USA	HYDROS	250-600	2511	Liq. Water	-

Solid Rocket Propulsion (SRP) systems
In this method, thrust is generated by burning a solid propellant in a combustion chamber and ejecting the resulting products of combustion through a nozzle as shown in **Fig. A10-7**.

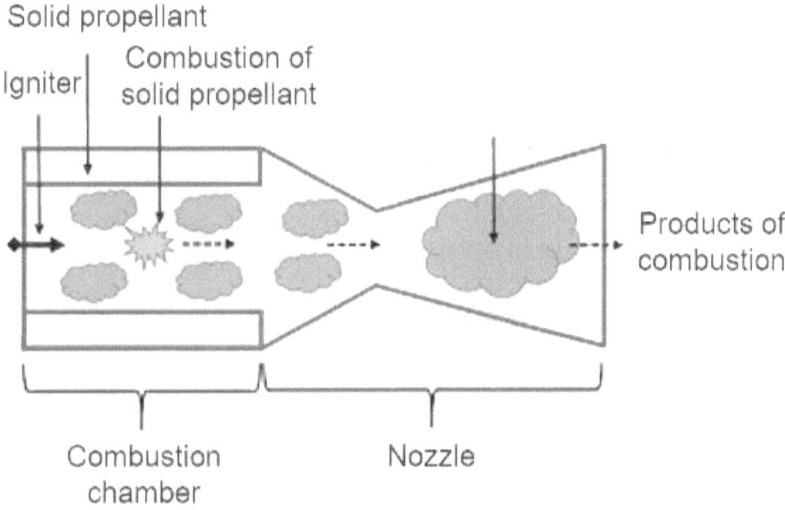

Solid propellant

Igniter

Combustion of solid propellant

Products of combustion

Combustion chamber

Nozzle

▲ **Fig. A10.7:** Solid Rocket Propulsion system

Though the system is simple to build and operate, solid propellant rockets offer only a one-shot use as the propellant cannot be replenished. An array of such rockets will have to be carried by the satellite to suit mission needs. A typical SRP micro-thruster array makes use of MEMS technology and comprises several laminated layers containing a combustion chamber, an igniter, a nozzle and a seal. The combustion chamber stores the solid propellant and the igniter section heats the propellant by means of a silicon or nichrome resistive heating element to initiate combustion. Some currently available solid rocket propulsion systems are shown in **Table. A10-3.**

▼ **Table. A10-3:** Examples of solid propulsion systems

Company	Engine	Thrust (mN)	I_{sp} (Ns/kg)	Propellant	Heritage
Orbital ATK, USA	STAR 4G	13	2642	Al & Ammonium Perchlorate	-
DSSP, USA	CAPS-3	-	2403-2550	HIPEP-501A	SPINSAT (57 kg) (Deployed from ISS in 2014)
DSSP, USA	CDM-1	76	2217	AP/HTPB	-

Resistojets:

In a resistojet, the propellant is passed through a heat exchanger where it is super-heated and ejected through an expansion nozzle as indicated in **Fig. A10-8.**

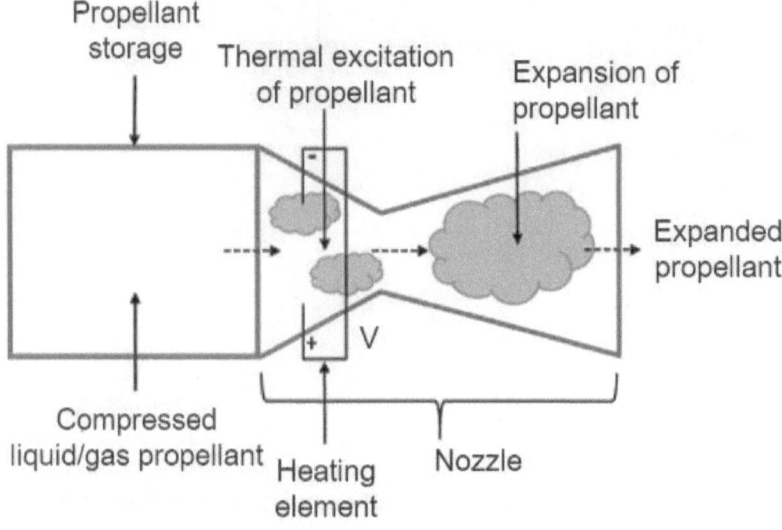

▲ **Fig. A10.8:** Resistojet

Laboratory experiments have shown exit temperatures of 600–1050° Centigrade for methanol and 300–1175° Centigrade for ammonia propellants. The working principle of a resistojet is similar to that of a cold gas propulsion system except that the propellant is heated before the expansion process. Exit velocities of micro cold gas propulsion systems range between 300–700 m/s, while those of micro resistojets are about 2.2 km/s. A major drawback of resistojets is that their performance (thrust and specific impulse denoted as Isp) is limited by the melting temperature of the heating element used.

Resistojets have been used on larger satellites and they have the potential for use on microsatellites.

Some currently available Resistojets are shown in **Table. A10-4.**

▼ **Table. A10.4:** Examples of resistojets

Company	Engine	Thrust (mN)	Power (W)	I_{sp} (Ns/kg)	Propellant	Heritage
SSTL, UK	LPR	18	30	471	Xe	NovaSAR-1 (430 kg; PSLV C42-2018)
VACCO Industries, USA	PUC	5.4	15	637	SO_2	-
VACCO Industries, USA	CHIPS	30	30	804	R134a, R236fa	-
Busek, USA	AMR	10	15	1471	R134a, R236fa	-
Busek, USA	FMMR	0.129	-	777	Water	-

Radio-frequency ion thruster:

These engines generate thrust by accelerating an ionised propellant (plasma) through an electrostatic grid. A typical radio frequency ion thruster is depicted in **Fig. A10-9**.

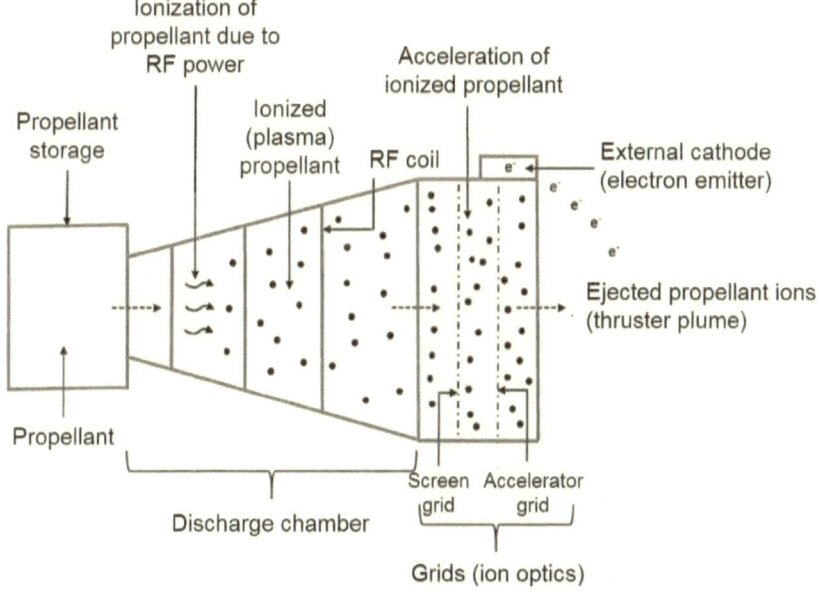

▲ **Fig. A10.9:** Radio frequency ion thruster

A stored propellant such as xenon or krypton is let into the discharge chamber. where it is ionised and becomes plasma by means of radio frequency (RF) power. The ionised propellant is then extracted from the discharge chamber and accelerated by a series of grids called screen and accelerator grids. The screen grid extracts propellant cations (for instance, Xe+, Kr+ ions) from the ionised plasma and directs them downstream towards the accelerating grid. A neutraliser cathode, present on the exterior of the thruster, provides electrons to neutralise the ionised propellant that is emitted from the thruster.

These thrusters have no flight heritage so far. Some of the systems available are shown in **Table. A10-5.**

▼ **Table. A10.5:** Examples of radio frequency ion thrusters

Company	Engine	Thrust (mN)	Power (W)	I_{sp} (Ns/kg)	Propellant	Heritage
Busek, USA	BIT-1	0.1-0.18	28	21090-31390	Xe, Iodine	-
Busek, USA	BIT-3	1.15	75	24525	Iodine	Will fly on Lunar Ice Cube (14 kg / 6U) (SLS Artemis 1 - 2020)
Airbus, Germany	RIT-μX	0.05-0.5	<50	2940-29400	Xe	-
Airbus, Germany	RIT 10 EVO	5,10,15	145	>18639 >29430 >31390	Xe	Available in 3 designs

HALL EFFECT THRUSTERS

Hall Thrusters are electrostatic devices that generate thrust by first ionising and then accelerating the propellant in mutually perpendicular electric and magnetic fields. Hall effect states that when an electric current is applied to a conductive material (propellant) placed in mutually perpendicular electric and magnetic fields, a potential difference is developed that is perpendicular to the applied electric and magnetic fields.

A Hall thruster includes propellant storage, discharge channel, external cathode, anodes and the magnetic field generator as shown in **Fig. A10-10.** The applied magnetic field is radial, while the accelerating electric field (acting from anode towards cathode) is axial. The ionised propellant is accelerated due to the strong electric field. They have high specific impulse but susceptible to erosion due to discharge plasma.

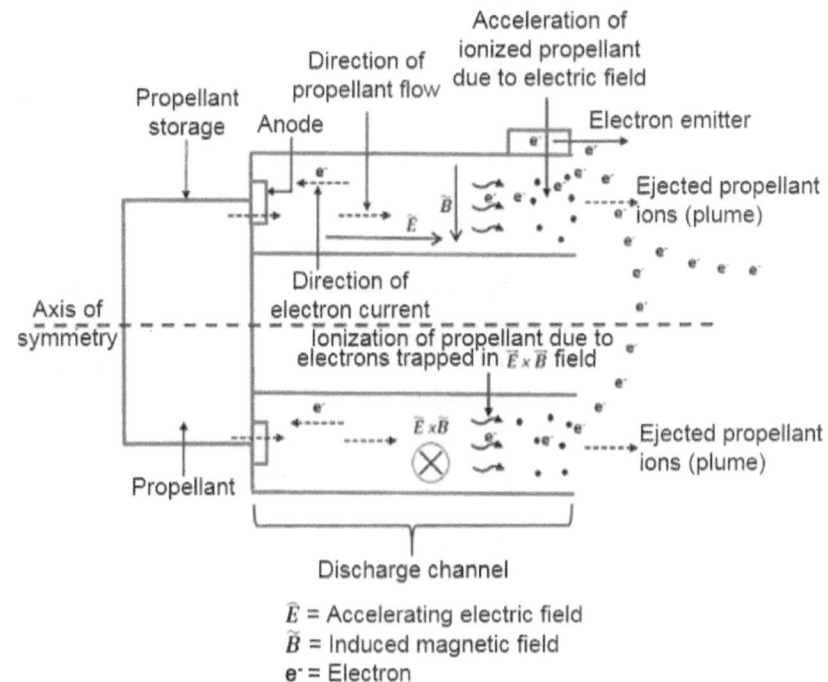

▲ **Fig. A10.10:** Hall effect thruster

Status of development of Hall effect thrusters is shown in **Table. A10-6**.

▼ **Table. A10.6:** Examples of hall effect thrusters

Company	Engine	Thrust (mN)	Power (W)	I_{sp} (Ns/kg)	Propellant	Heritage
Busek, USA	BHT-200	12.8	200	13636	Xe, I, Kr	TacSat-2 (370 kg; 2006) FalconSat-5 (180 kg; 2010)
Sitael, Italy	HT-100	10	100	10790	Xe, Kr	-
Sitael, Italy	HT-400	50	100	17167	Xe	-
MIT, USA	MHT-9	20-50	30-200	2943-14725	-	-

ELECTRO-SPRAY THRUSTERS

An electro-spray thruster is a plasma-free electric propulsion system that works on the principle of electrostatic extraction and acceleration of charged particles (ions) from a liquid propellant surface to produce thrust. The working mechanism is based on a process by which the conductive liquid surface of propellant is deformed into a sharp cone-shaped meniscus called the Taylor cone. When a certain threshold of the electric potential is surpassed, ions are extracted from the cone's apex. The propellant is an ionic liquid. Positive or negative ions produced are accelerated to generate either a positive or negative ion beam thereby eliminating the need for an external cathode to neutralise the ejected ions unlike in ion and Hall thrusters.

Major components of an electrospray thruster comprise propellant storage, emitter and extractor electrode as shown in **Fig. A10-11**. The performance of an electrospray thruster can be varied by changing the current passed through the emitter and the extractor electrodes.

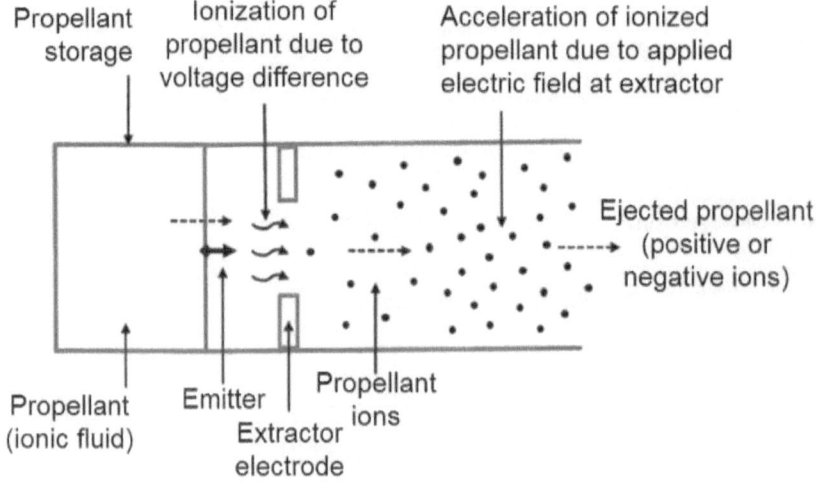

▲ **Fig. A10.11:** Electrospray thruster

An individual electrospray emitter operates in the milli-watt range and generates thrust in the order of micro-Newtons; therefore an array of emitters is required to form the thruster that can yield the desired thrust. Electro-spray thrusters have been used on a CubeSat as well as on a bigger satellite. Some currently available electrospray thrusters are shown in **Table. A10-7**.

▼ **Table. A10.7:** Examples of electrospray thrusters

Company	Engine	Thrust (mN)	Power (W)	I_{sp} (Ns/kg)	Propellant	Heritage
MIT, USA	S-iEPS	0.1	1.5	11772	Ionic fluid	AeroCube-8 (1.5U; 2015)
Accion Systems, USA	TILE5000	1.5	30	17658	Ionic fluid	-
Busek, USA	BET-1mN	0.7	<9	7848	Ionic liquid	-
Busek, USA	BET-100	0.005-0.1	5.5	17658	Ionic fluid	-

Pulsed plasma thruster

A Pulsed plasma thruster (PPT) operates by creating a pulsed, high-current discharge across the exposed surface of a solid insulator (e.g., Teflon) that serves as a propellant. An arc discharge vaporises the propellant from its surface thereby ionising and accelerating the propellant to high speeds. A typical pulsed plasma thruster is depicted in **Fig. A10-12**.

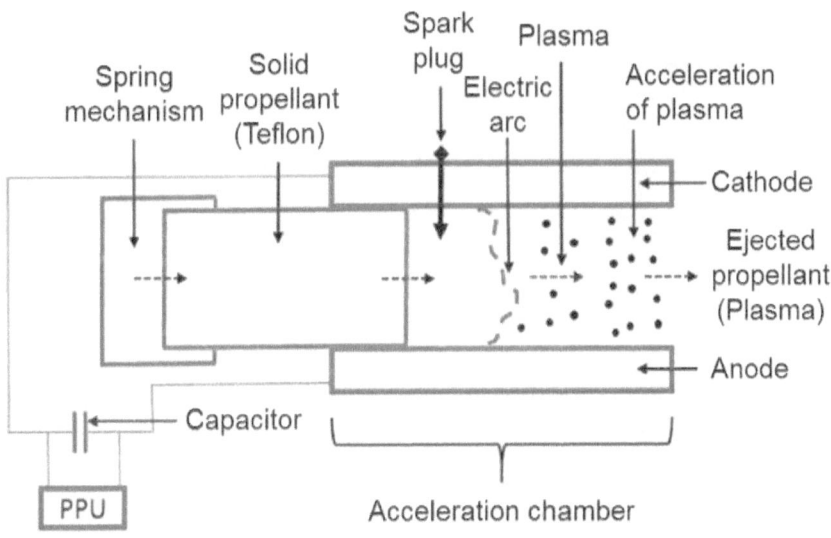

▲ **Fig. A10.12:** Pulsed plasma thruster

A current pulse is sent out to electrodes lasting few microseconds, which is generally driven by a capacitor that is charged and discharged approximately once every second. The thruster consists of a spring-loaded mechanism, propellant, capacitor, anode, cathode, acceleration chamber and a spark plug. A spring feeds the propellant between the two electrodes and the spark plug is simultaneously fired to raise the electrical conductivity of the acceleration chamber. The electric current from a power processing unit (PPU) flows to the electrodes through the capacitor and then into the arc thereby completing a current loop and simultaneously generating a magnetic field. The ionised plasma produced by the vaporised propellant is then accelerated due to the Lorentz force generated by the electric arc and the induced magnetic field.

Modified PPT technologies such as planar pulsed inductive thruster (PIT) and vacuum arc thrusters (VAT) are also under development. These differ in the way the ionised plasma is produced. Some currently available PPTs and VATs are shown in **Table. 10-8**.

▼ **Table. A10.8:** Examples of pulsed plasma thrusters

Company	Engine	Thrust (mN)	Power (W)	I_{sp} (Ns/kg)	Propellant	Heritage
Primes, USA	EO-1-PPT	0.14	12.5	11281	Teflon (PPT)	Dawgstar (13 kg; Not flown)
Busek, USA	MPACS	0.144	<10	8142	Teflon (PPT)	FalconSat-3 (54.3 kg; Atlas 5-2007)
GWU, USA	μCAT	0.001-0.02	<10	29430	Nickel (VAT)	BRICSat-P (1.5U/1.9 kg; Atlas 5-2015)
Wurzburg University, Germany	UWE4	0.002-0.01	0.5-2	10790	Titanium, Tungsten (VAT)	-

Solar Sails

A solar sail is a form of propellant-less spacecraft propulsion system that generates thrust by means of momentum exchange with the incoming solar radiation. Solar sails have a flat surface and are usually made of thin reflective material supported by a lightweight deployable structure as shown in **Fig. A10-13**. As they do not use a propellant, solar sails by definition, possess infinite specific impulse. However, the main drawback of a solar sail is very low thrust levels resulting in the need for large surfaces to be deployed and a long time to gain appreciable momentum change.

▲ **Fig. A10.13:** Solar sail (Courtesy: Space.com)

In June 2015, the short-lived LightSail-1, a 3U CubeSat deployed its solar sail in the earth orbit. LightSail-2, which was launched on June 25, 2019, has demonstrated controlled solar sailing in the earth orbit.

Current technology maturity status

Fig. A10-14 shows the current maturity status of the various micro-propulsion technologies discussed so far. Due to rapid developments in MEMS techniques and ongoing research in advanced thruster systems, propulsion systems will find application in microsatellites very soon.

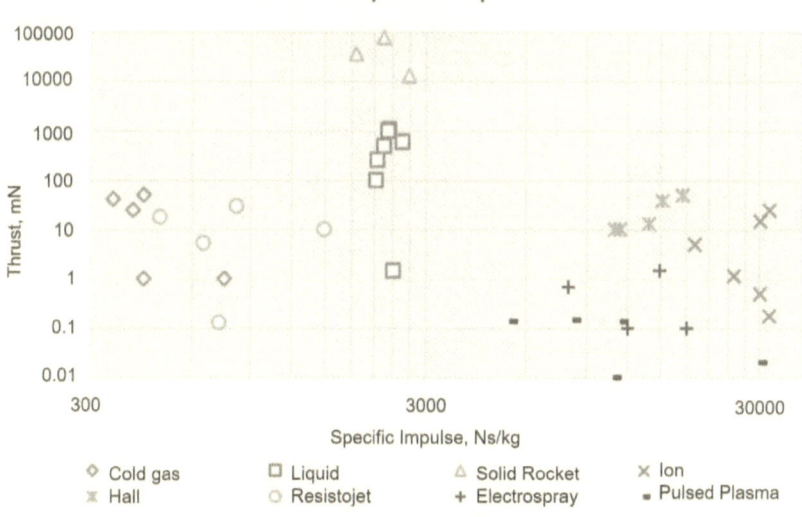

▲ **Fig. A10.14:** Current technology maturity status for propulsion systems

ANNEXURE - A11
COMMERCIAL ADCS HARDWARE

Details of various hardware units used in ADCS of a Nanosatellite that are commercially available are given below:

Reaction wheels

Parameter	Unit	MAI	Cubespace	Blue Canyon
Momentum storage	mNms	11.076	1.77	15
Maximum torque	mNm	0.635	0.23	4
Mass	Gms	110	60	130
Power	mW	850	150	1
Dimension	Mm	33x33x38	28x28x26	43x43x18

Inertial measurement unit (IMU)

Parameter	Units	Honeywell	Analog Devices
Measurement range	Deg/sec	+1440	+450
Bias stability	Deg/hr (1 sigma)	1	0.8
Random walk	Deg/Sqrt-Hr (1 sigma)	0.125	0.09
Mass	Gms	160	43
Power	W	<3	0.3
Dimension	Mm		45 x 50 x 15

Star tracker

Parameter	Unit	NSTR-1	MAI-SS	SINCLAIR-Interplanetary
Field of view	Deg	15x12	-	8x8
Attitude accuracy	Arc-sec	7,7,70	4,4,27	5,5,55
Update rate	Hz	10	4	2
Mass	Gms	245	282	185

Sun sensor

Parameter	Unit	NCSS-SA05	NanoSSOC-A60
Field of view	Deg	114	120
Accuracy	Deg	0.5	0.5
Mass	Gms	5	3.7
Power	mW	<50	<10

NSS Cubesat ADCS Board

A CubeSat's ADCS measures orbit position, absolute attitude, satellite rates and can also control the orientation of the satellite through either magnetic actuation or wheel-based solutions. The new space ADCS board is a single PC104 board (15mm height) that physically integrates three high accuracy sun sensors, a magnetometer, MEMS gyros, a stellar gyro and two magnetic torquer rods. Additionally, external interfaces allow for a further three sun sensors, the Z-axis magnetic torquer and momentum or reaction wheels.

ADCS board specification

	Unit	Magnetic Control	Adding gyro and wheel	Adding GPS Receiver
Pointing knowledge	Deg	2	2	1
Control accuracy	Deg	10	3	1
Magnetic moment	Amp-meter square	0.2	0.2	0.2
Position accuracy	M	-	-	50
Velocity accuracy	cm/sec	-	-	25
Dimension	mm	96x96x15	96x96x15	96x96x15
Mass	gms	<200	<400	<450
Power	W	1.8	2	3

▲ **Fig. A11.1:** ADCS Board (Courtesy: New Space)

ANNEXURE - A12
ADCS CONFIGURATION IN INDIAN NANOSATELLITES

Few examples of Nanosatellites developed by educational institutions in India and their ADCS system are given below:

Satellite	Mass (kg)	Type of stabilisation	Sensors	Actuators	ADCS modes
PISAT (PES Univ., Bangalore)	5	Three axes stabilised	Magnetometer, gyroscope, Sun sensor	Magnetic Torquers (3)	De-tumbling normal mode(imaging), safe mode
JUGUNU (IIT – Kanpur)	3	Three axes stabilised	Magnetometer, gyroscope, sun sensor, GPS	Reaction Wheels(3) Magnetic Torquers (2)	De-tumbling, fine pointing imaging mode, momentum dumping, safe mode
PRATHAM (IIT – Mumbai)	10	Three axes stabilised	Magnetometer, gyroscope, sun sensor, GPS	Magnetic Torquers	De-tumbling normal mode(imaging), safe mode
SWAYAM Engg. College, Pune)	1	Passive Magnetic stabilisation	MEMS gyro	Permanent magnets and hysteresis rods	Rate damp, antenna aligning along earth's magnetic field

ANNEXURE – A13
RELIABILITY PREDICTION MODELS

Mathematical models may either be deterministic or probabilistic in nature. In deterministic models, results can be predicted exactly. In contrast, a probabilistic model is one in which the results cannot be determined as exactly as in the deterministic model but can only be described in terms of probability or a probability distribution function.

The discipline of reliability engineering is based upon the use of deterministic and probabilistic or stochastic models. Some of the reasons supporting the use of probabilistic models are:

1. Accurate quantification of the variables that contribute to the failure of all of the electronic components in order to develop exact deterministic failure models

2. Probabilistic models, when applied to large samples, tend to smooth out individual variations. Thus, the final average result is simple and accurate enough for engineering analysis and design

Reliability parameters are defined in probabilistic terms; therefore, probabilistic parameters such as random variables, density functions and distribution functions are utilised in the development of reliability theory and practice.

STATISTICAL DISTRIBUTIONS USED IN RELIABILITY MODELS

Many standard statistical distributions may be used to model various reliability parameters. However, a relatively small number of statistical distributions satisfy most reliability needs. The particular distribution used depends upon the nature of data. The most appropriate model for a particular application may be decided either empirically or theoretically or by a combination of both approaches.

A distribution may be chosen empirically by fitting to data. Simple graphical methods have been developed for fitting distributions. Following is a short summary of some of the distributions commonly used in reliability analysis, the criteria for their use and some examples of this application.

CONTINUOUS DISTRIBUTIONS: NORMAL (OR GAUSSIAN) DISTRIBUTION

There are two principal reliability applications of normal distribution. The first deals with the analysis of items that exhibit failure due to wear such as mechanical devices. Frequently, the wear-out failure distribution is sufficiently close to normal that the use of this distribution for predicting or assessing reliability is valid.

The second application deals with the analysis of manufactured items and their ability to meet specifications. Use of the normal distribution in this application is based upon the central limit theorem. It states that the sum of a large number of identically distributed random variables, each with a finite mean (u) and a standard deviation is normally distributed. Thus, the variations in parameters of electronic components due to manufacturing reasons are considered to be normally distributed.

Log-normal distribution

If the natural logarithm of a function is found to be distributed normally, then the function is said to be log-normal. Physical examples of the log-normal distribution are the fatigue life of certain types of mechanical components. Semiconductor failures may also frequently follow a log-normal distribution.

Exponential distribution

This is the most important distribution in the reliability work. It is used almost exclusively for reliability prediction of electronic equipment. It is completely defined by a single parameter namely the failure rate (λ). Some major advantages of the exponential distribution:
- Single, easily estimated parameter (λ)
- Mathematically very tractable
- Wide applicability
- It is additive (sum of a number of independent exponentially distributed variables is exponentially distributed)

Specific applications include:
1. Items whose failure rate does not change significantly with age
2. Complex repairable equipment without excessive amounts of redundancy
3. Equipment from which the infant mortality or early failures have been eliminated. This is done by putting the equipment in continuous operation termed as "burn-in" for the equipment for some time period under normal operating condition or by the accelerated condition of higher temperature

The hazard rate for the exponential distribution is constant and equal to λ. This property is the basis for MIL-HDBK-217's comprehensive tabulation of failure rates for various components. They are based on the assumption of a constant failure rate. The underlying assumption for the use of the exponential distribution is randomness. The exponential distribution may also be recognised as a special case of the gamma distribution.

Gamma distribution

Gamma distribution is used in reliability analysis for those cases where partial failures can exist, that is when a given number of partial failures must occur before an item fails, for example, redundant systems. Gamma distribution can be used to describe either an increasing or a decreasing hazard (failure) rate.

Weibull distribution

The Weibull distribution, named after the Swedish investigator of metal fatigue problems, W. Weibull, is especially useful in reliability work. It is a very general distribution which by adjustment of the distribution parameters can be made to model a wide range of life distribution characteristics for different classes of items. One of the versions of the failure density function is

$$\text{Where} \quad f(T) = \frac{\beta}{\eta}\left(\frac{T-\gamma}{\eta}\right)^{\beta-1} e^{-\left(\frac{T-\gamma}{\eta}\right)^{\beta}}$$

β= shape parameter
η= scale parameter or characteristic life
γ= minimum life

In most practical reliability situations, $\gamma=0$. Depending upon the value of β, the Weibull distribution function can also take the form of the following distributions:

Beta value	Distribution type	Hazard rate
< 1	Gamma	Decreasing
1	Exponential	Constant
2	Log - normal	Increasing/decreasing
3.5	Normal (approximately)	Increasing

By appropriate choice of the value of β, the Weibull distribution can be used to describe each of the three regions of the bathtub curve, infant mortality, useful life and wear out.

Discrete distribution - binominal distribution

The binominal distribution is very useful in both reliability and quality assurance. It is used where there are only two possible outcomes such as success or failure and probability remains the same for all trials.

Poisson distribution

This distribution is used quite frequently in reliability analysis. It can be considered as an extension of the binominal distribution when n is infinite. In fact, it is used to approximate the binominal distribution when $n > 20$ and $P \leq 0.05$.

If events are Poisson distributed, they occur at a constant average rate and the number of events occurring in any time interval is independent of the number of events occurring in any other time interval.

Bayesian statistics in reliability analysis

Bayesian statistics are being used increasingly in reliability analysis. The advantage of Bayesian statistics is that it allows prior information (for example, predictions, test results, engineering judgment and so on) to be combined with more recent information, such as test or field data, in order to arrive at a prediction or assessment of reliability based upon a combination of all available data.

System reliability analysis

The first step in the reliability engineering process is to specify the numerical requirement. A reliability requirement must be specified quantitatively to be meaningful. Four commonly used ways of quantitatively defining a reliability requirement are mean-time-between-failures, probability of survival, probability of success and failures per million hours.

Mean-time-between-failures (MTBF)

MTBF is applicable for long life equipment and systems where the planned mission lengths are typically short relative to the specified MTBF.

Probability of survival, R (t)

Applicable for defining reliability when a high reliability is required during the mission period but an MTBF greater than the mission period is of little consequence.

Probability of success, P(s)

Probability of success is independent of time, used for specifying the reliability of one-shot devices, such as the flight reliability of missiles and so on.

Failures per million hours (FPMH) (λ)

Failure rate (λ) over a specified period of time is used to specify the reliability of components, units and assemblies whose reliability for the time period of interest approaches unity. For equipment exhibiting an exponential failure distribution, it is the reciprocal of the MTBF.

ANNEXURE - A14
USE OF COTS COMPONENTS IN NANOSATELLITE

Electronic components used in satellites are available in different quality grades:
- Space qualified or class-S grade
- Qualified to military standards (**MIL-883B grade**)
- Industrial grade
- Commercial-off-the-shelf grade (**COTS**)

These quality grades are derived from their operating temperature range as shown below, in addition to several other factors:

Grade	Operating temp range
Space qualified, Class-S	- 55 to + 125 deg C
Military grade (MIL-883B)	- 55 to + 125 deg C
Industrial grade	- 40 to + 85 deg C
COTS	**0 to + 70 deg C**
Automotive grade	- 40 to +105 deg C

For operational satellites with long mission life (>10 years), the reliability factor should be as high as possible which, in turn, is derived from the reliability of the components used in the satellite. Space Agencies such as NASA and ESA (European Space Agency) and Defence Departments have established the reliability of electronic components by conducting several tests on them and evolved a set of standards to which all manufacturers need to comply. Some of these tests include:
- Electrical 'burn-in' tests for 168 hours
- Constant acceleration test
- Temperature cycling test
- Gross/Fine leak test
- Bond pull/die shear test
- Particle impact noise detection (PIND) test
- Exposure to radiation test

- Steady-state life test
- Internal/external visual inspection
- Destructive physical analysis (DPA)

By conducting these tests, one can weed out all defective components at the manufacturers' end. Based on the design, defective raw materials used, manufacturing process, workmanship and operating conditions, the life of electronic components follows the 'bathtub' curve as shown in **Fig. A14-1.** Defective components fail during the early stages of testing (infant mortality) and stressed components fail due to wear out. Hence, components which withstand infant mortality can be safely used in satellite applications. The predicted failure of electronic components is generally expressed in terms of 'mean-time-between-failure' (MTBF), which is defined as the average amount of time a device functions before it fails. This time refers to only operational time and does not include non-use periods.

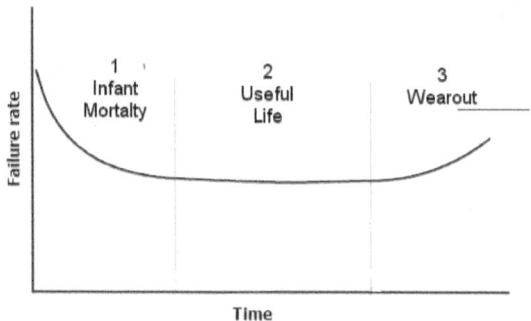

▲ **Fig. A14.1:** Bathtub curve for the reliability of electronic components

Selection of the quality of components for a Nanosatellite project directly impacts two factors:
- Reliability of the satellite
- Cost of the satellite

Before selecting the quality, one has to address the following questions:
- ✓ How critical is the mission goal? **Very critical or experimental**
- ✓ How long is the mission life? **>2 yrs or <2 years**
- ✓ What is the project schedule? **<2 yrs or >3-4 years**
- ✓ What is the budget allocation? **<Rs. 1 crore or >Rs. 5 crore**
- ✓ Programme plan? **One time mission or series of satellites**

In most of the cases, particularly for educational institutions with limited resources, the outcome of this trade-off analysis is to use COTS components. Hence a brief description of the issues and precautions to be taken for using COTS electronic components is given below.

Major differences between space grade and COTS components are summarised in **Table. A14-1** below. Basically, COTS components are manufactured for the auto and mobile phone industry and easily available at low cost.

▼ **Table. A14.1**

Space grade	COTS
Space grade components are costly and long lead for procurement	COTS are cheaper and readily available (short lead times)
Qualified and screened by the supplier at source	Up-screening by users to enhance confidence
Good traceability with lot/date codes and test reports	Traceability records with the manufacturer only
Radiation hardened by the process	Radiation hardening not guaranteed
Supplied with ceramic packages	Supplied in plastic encapsulated micro-circuits
Limited manufacturers	Several manufacturers

As they do not supply test data records with components, it is a normal practice to conduct screening tests on a few samples after procurement to gain confidence about their quality. These screening tests include:

- Electrical functional tests at 25°C, 70°C and 0°C
- Burn-in at 125°C for 168 hrs (sample devices) – check for drift in parameters
- Thermal cycling - 40°C to + 85°C (either component or wired PCB level)
- Highly accelerated stress test 80°C /85% RH for 96 hrs
- Radiation hardness assurance test (total ionising dose, high energy protons) on sample devices

Following guidelines may be followed while procuring the COTS components for Nanosatellite:

I. As far as possible, procure COTS from reliable vendors, who supply space/ MIL grade components also with latest date codes
II. Avoid buying from agents/unauthorised stockists, who may supply spurious components
III. Always procure with buffer stock to meet any contingencies/failures
IV. Since COTS are supplied in plastic packages, ensure the parts are procured in ESD safe bags to prevent corrosion and moisture ingress during shipment and storage

V. After receiving COTS components, they shall be subjected to incoming inspection including visual inspection, electrical testing, review of data, followed with destructive physical analysis (DPA) on samples

One of the major causes for failure of COTS components in orbit is due to space radiation of high energy protons. Since COTS components are NOT radiation hardened by the process, it is important to take precautions to minimise the radiation effects. Some of them are:

- Include a watch-dog-timer to monitor and restore an unintended state of the processor and related devices
- Monitoring and 'power ON' resetting for 'locked-up' communication devices
- Circumventing higher circuit currents, which invariably occur during single event latch-up conditions using suitable over current protection circuit. Hardware and associated software to watch for increased current levels and to provide needed power reset
- Software error detection and correction (EDAC) provisions for memory device malfunctions by routinely scrubbing and verifying multiple copies of the data
- Preference of allowing 'power OFF' conditions for devices/circuits, which are not required during certain operations
- Implementation of 'triple modular redundancy' (TMR) for critical functions, where logics are replicated and usable output is selected by majority-voting logic (MVL)
- Schemes to ruggedise protocols for data communication where the common data bus is shared by multiple peripherals. Suitable software that periodically polls 'unintended' conditions could be considered

Some of the other precautions to be taken during fabrication of electronic hardware using COTS components are:

- Since COTS are not hermetically sealed, they should be stored and handled with care (in ESD safe bags)
- Electrostatic discharge (ESD) failures to be avoided by using wrist straps or ESD mats
- Pure tin in soldering should not be used to avoid the growth of 'whiskers' at solder joints
- Shielding with tantalum sheet on VLSI devices may be implemented for sensitive and critical devices
- During the design phase, proper guidelines for 'derating' the operating conditions for components shall be established and followed

ANNEXURE - A15
PRINCIPLES OF SOFTWARE TESTING

The seven basic principles of software testing are:

1) Exhaustive testing is not possible

Yes! Exhaustive testing is not possible. Instead, we need the optimal amount of testing based on the risk assessment of the application. And the million-dollar question is, how do you determine this risk?

2) Defect clustering

This states that a small number of modules contain most of the defects detected. This is the application of the Pareto principle to software testing: approximately 80 per cent of the defects are found in 20 per cent of the modules.

3) Pesticide paradox

Repetitive use of the same pesticide to eradicate pests in farming results in the pests developing resistance to the treatment rendering the pesticide ineffective. This applies to software testing also. Conducting the same set of tests repetitively will not help discover new defects. To overcome this, the test cases need to be regularly reviewed and revised, adding new and different test cases to help find more defects. That is testers cannot simply depend on existing test techniques. Testers must look out continually to improve the existing methods to make testing more effective.

4) Testing shows the presence of defects

The testing principle states that testing talks about the presence of defects and not the absence. That is to say, software testing reduces the probability of undiscovered defects remaining in the software but even if no defects are found, it is not a proof of correctness.

5) Absence of error

A software which is 99 per cent bug-free may be still unusable. This can be the case if the system is tested thoroughly for the wrong requirement. Software testing is not merely finding defects but also to check that software addresses the end-use as per requirements. Absence of error is a fallacy i.e., finding and fixing defects does not help if the system built is unusable and does not fulfil the user's needs and requirements.

6) Early testing

Testing should start as early as possible in the software development life cycle. This is to capture any defects in the requirements or design phase in the early stages. It is much cheaper to fix a defect in the early stages of testing. But how early one should start testing? It is recommended that one should start testing for a bug the moment requirements are defined.

7) Testing is context-dependent

Testing is context-dependent, which basically means that the way you test an e-commerce site will be different from the way you test a commercial off the shelf application. All the developed software are not identical. You might use a different approach, methodologies, techniques and types of testing depending upon the application type.

ANNEXURE - A16
ELECTRICAL GROUNDING SCHEMES

There are two types of grounding schemes generally implemented in a satellite's electrical configuration: single-point grounding and multiple-point grounding. In the single-point grounding scheme, shown in **Fig. A16.1,** all circuit commons are grounded by means of connecting them to one single point on the structure also known as 'satellite ground reference point (SGRP)'. Note the isolation of grounds (circuit commons) between assemblies, so that there is one and only one DC ground reference path for each assembly. This is sometimes called a "star" ground because all ground wires branch out from the central point of the star. However, inductances of long wires and higher frequencies can negate the adequacy of this scheme such that the assembly may no longer have a zero potential reference with respect to chassis.

Isolation between assemblies.

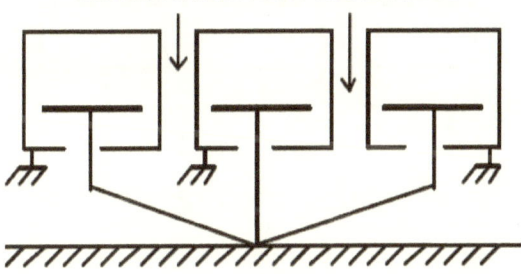

▲ **Fig. A16.1:** Single-point "star" ground (Courtesy: NASA-HDBK-4001)

Figure A16.2 shows a multiple-point (multi-point or multi-path) grounding arrangement. In this case, each circuit common is grounded directly to the chassis and also grounded indirectly to the chassis via the connections to the other assemblies. This is typically followed for radio frequency (RF) sub-systems but should not be used for video or other signals containing low frequencies (less than roughly 1 MHz).

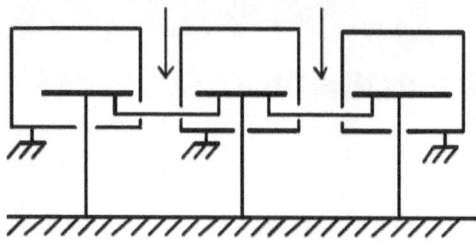

No isolation between assemblies.

▲ **Fig. A16.2:** Multiple-point ground (Courtesy: NASA-HDBK-4001)

Fig. A16.3 shows a better chassis referenced grounding scheme where each assembly has one and only one path to the chassis (the zero voltage reference) and there are no deliberate structure currents. Compared to the single point grounding scheme, each ground reference wire is short providing minimum AC impedance between each circuit common and the chassis. The main advantage of this scheme is that each electronic item has one and only one path to chassis and there is no deliberate chassis current. Also, all sub-systems have a common DC voltage reference potential (the interconnected structure).

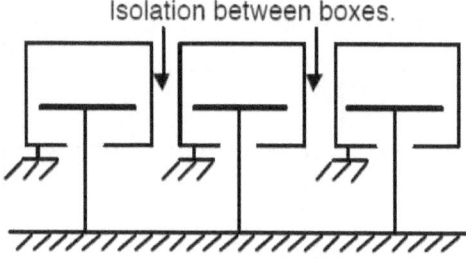

Isolation between boxes.

▲ **Fig. A16.3:** Sub-system level grounding scheme (Courtesy: NASA-HDBK-4001)

ANNEXURE - A17
COMMERCIAL INDUSTRIES - SMALL SATELLITE SYSTEMS

This compilation has attempted to highlight small satellite systems available in the commercial space industry and usable by entrepreneurs. It is not exhaustive and is based on published literature. Available information on independent verification carried out by NASA and other leading agencies have also been considered. The offered systems and technology are cited to have reasonable heritage with several successful missions.

It is difficult to obtain details on valid confirmation of quality data, flight performance and pedigree for most of the systems. For Nanosatellite entrepreneurs, it is preferable to adopt suitable screening and validation methods to gain confidence on their usage for envisaged application.

Small Spacecraft Bus	
➤ Integrated spacecraft platforms ➤ Off-the-shelf for rapid integration and delivery ➤ Complete CubeSat buses ➤ turnkey customised platforms	1. Surrey Technology Ltd, Europe 2. Tyvak Nanosatellite Systems Inc 3. Blue Canyon Technologies, USA 4. Gom Space, Europe 5. Millennium Space Systems, USA 6. Spaceflight Industries, USA 7. Astro-und Feinwerktechnik Adlershof GmbH 8. Berlin Space Technologies GmbH
UHF/VHF Communication Systems	
➤ Transceivers ➤ High gain antennae ➤ Deployable antenna systems	1. BitBeam Inc. 2. Clyde Space Ltd. 3. Gom Space 4. L3 Communications Inc. 5. Innovative Solutions In Space (ISIS)
S-Band Communication Systems	
➤ Transponders ➤ TDRSS compatible systems ➤ Patch and quadric filar antenna systems	1. Clyde Space Ltd 2. Haigh-Farr Inc 3. Innovative Solutions In Space (ISIS)

X-Band Data Transmission System

▷ MMIC based transmitters	1. Surrey Technology
▷ Patch and high gain antenna	2. Antenna Development Corporation
▷ Pointing mechanisms	

Power System

▷ Solar array (ITJ/UTJ cell-based	1. Clyde Space Ltd
▷ Power storage battery (Li-ion cell-based	2. Gom Space, Denmark
▷ Power management and distribution	3. Blue Canyon Technologies, USA
▷ Power converters	4. Surrey Technology
	5. Stras Space

Attitude Determination and Control System

▷ ADCS testbeds	1. Berlin Space Technologies
▷ Air bearing platforms with sun simulators and Helmholtz cage	2. Surrey Technology
	3. Space Micro
▷ Star trackers	4. New Space Systems
▷ Sun sensors	5. Analog Devices
▷ Magnetometers	
▷ Gyroscopes	

General

▷ CubeSat (form factor) Structures	1. Innovative Solutions In Space (ISIS)
▷ Orbit deployers	2. Texas Spacecraft Laboratory
▷ Risk analysis and management	

"Quintessence of Nano Satellite Technogy"

BOOK REVIEW

Space and Satellites used to be buzz words beyond the understanding of common men. No more - the evolution of Small and Nano Satellites has changed the paradigm and Nano Satellites are being designed and produced by students in relatively small colleges and universities. A new era of multidisciplinary innovations has opened up in which students of different disciplines learn to work together, evolve themselves as innovative problem solvers, not necessarily in space but in any multidisciplinary environment they may be put into.

This book, "Quintessence of Nano satellite Technology" by Planet Aerospace is to be seen in this context. It introduces the concept of small, mini, micro and Nano satellites and then gets into the basics of individual disciplines describing the principles of individual technology and design. Outlining the ideas of mission design as overall mission including orbital mechanics and how to design a satellite for specific objectives, design of primary systems like communication with satellite, management of power and thermal systems, design of attitude and orbit control system including interdependence of these systems and finally integration and Quality system in Space systems, that makes space systems different from terrestrial technology.

This book written by stalwarts of ISRO will prove to be an enabler for colleges and universities to take up the development of Nano Satellites, that could be launched by opportunities provided by ISRO and other international agencies. Hope, this will become a milestone in boosting Nano Satellite activities and demystifying space.

Dr. P. S Goel
Former Secretary, Ministry of Earth Sciences
And Director, ISRO Satellite Centre

www.ingramcontent.com/pod-product-compliance
Lightning Source LLC
Chambersburg PA
CBHW030942240526
45463CB00016B/1208